*f*P

THE RAVAGING TIDE

Strange Weather, Future Katrinas, and the Coming Death of America's Coastal Cities

MIKE TIDWELL

FREE PRESS

NEW YORK   LONDON   TORONTO   SYDNEY

FREE PRESS
A Division of Simon & Schuster, Inc.
1230 Avenue of the Americas
New York, NY 10020

First Free Press trade paperback edition 2007

FREE PRESS and colophon are trademarks of Simon & Schuster, Inc.

For information about special discounts for bulk purchases, please contact Simon & Schuster Special Sales at 1-800-456-6798 or business@simonandschuster.com

Designed by Davina Mock

Manufactured in the United States of America

1   3   5   7   9   10   8   6   4   2

The Library of Congress has cataloged the hardcover edition as follows:
Tidwell, Mike.
The ravaging tide : strange weather, future Katrinas, and the coming death of America's coastal cities / Mike Tidwell.
p. cm.
Includes bibliographical references and index.
1. Climatic changes—Forecasting. 2. Weather forecasting—United States.
3. Climatic changes—United States. 4. Climatic changes—Social aspects.
5. Global environmental change. 6. Climate and civilization—United States.
7. Human beings—Effect of environment on—United States. I. Title.
QC981.8.C5T53 2006
363.738'74—dc22   2006041332
ISBN-13: 978-0-7432-9470-6
ISBN-10:   0-7432-9470-X
ISBN-13: 978-0-7432-9471-3 (Pbk)
ISBN-10:   0-7432-9471-8 (Pbk)

Acknowledgments

MANY THANKS first to Anne Havemann, whose tenacious research kept this book project moving forward on a tight schedule, and whose cheerful encouragement kept me sane during the writing. I'm honored, Anne, to call you a colleague and a friend.

Thanks also to the rest of the staff at the Chesapeake Climate Action Network and the U.S. Climate Emergency Council. I couldn't have undertaken this project without the support and patience of all of you.

I owe a debt of gratitude to Nick Varchaver and Frank Reiss for commenting on early drafts of this book, and to my dear old friend Mark Davis at the Coalition to Restore Coastal Louisiana for his constant guidance and insights. Mark fought heroically for years trying to prevent a storm like Katrina from happening, and he now speaks the truth like no one else as that state attempts to rebuild.

And very warm thanks to Beth, who took care of me more than anyone else during the long months of this project; and to my son, Sasha, the light of the world, who is the ultimate reason I do the work I do.

Finally, a huge thanks to my literary agent Jennifer Lyons who single-handedly rescued my writing career several years ago and is thus the main reason you're holding this book right now. Thank you, Jennifer, for believing in me and all my causes.

For Anne, Claire, Diana, Gary, Josh, Matt, and Ted

Contents

Introduction

IT WAS A SHORT PHONE CALL, lasting only a few minutes, but it formally launched the largest displacement of American citizens since the Civil War. On Saturday, August 27, 2005, Max Mayfield, director of the National Hurricane Center, told New Orleans mayor Ray Nagin that Hurricane Katrina was the "worst case" storm everyone had feared for decades. It was headed right for New Orleans with the energy of a ten-megaton nuclear bomb exploding every twenty minutes.

Within hours, Nagin had ordered the first mandatory evacuation in the city's three-hundred-year history. Over the next two days a staggering 1.3 million people would abandon the city and much of south Louisiana. So many cars headed north, full of people and pets and valuables, that satellite cameras captured the bumper-to-bumper interstate crawl from outer space.

In New Orleans, every rental car, every U-Haul van and truck, was gone. People walked, hitchhiked, hot-wired postal vehicles. They took flatboats up the Mississippi River. Amtrak and Greyhound sent their last cars and buses rolling north, east, and west—anywhere away from the storm. Prisoners were hustled off in chains. Hospital patients who could be moved were evacuated—babies in incubators, psychiatric patients strapped to gurneys. Drivers out of gas on clogged highways drilled holes in the gas tanks of abandoned cars for fuel to keep moving.

The human tidal wave crashed first into Louisiana towns just to the north. Baton Rouge, the somewhat somnolent state capital, doubled in size almost overnight, taking on 200,000 newcomers and becoming the largest city in the state just as New Orleans shrank to nothing. Hotels everywhere were booked solid. Extended families of up to forty people crammed into three-bedroom homes, with sleeping bags spread across hallways and kitchens, and water running nonstop from showers, washing machines, and flushing toilets.

And still they came, hundreds of thousands more refugees, arriving just ahead or after Katrina's harrowing landfall. Makeshift shelters sprang up across Louisiana and neighboring states and as far away as Nevada and Washington, D.C. Within days, Baton Rouge's modest airport was the second busiest in America, with passengers accepting any flight anywhere away from the storm and its aftermath, scattering themselves across America. By Sunday, September 4, the last fleeing inhabitants of New Orleans—the poorest and most desperate people, abandoned on overpasses and littered sidewalks—were finally bused by the thousands to Houston's Astrodome and convention center.

A week after it started, the retreat was at last complete. It had occurred on a scale no one could have imagined. Over one million people displaced in Louisiana alone. A vast section of American real estate lay broken and eerily, impossibly, empty. The return date for evacuees was wholly uncertain. Many would never return.

Yet as difficult and chaotic and disruptive as the Katrina evacuation was—broadcast nightly in horrifying detail to the world—there's one crucial element I'm sure most Americans have failed to appreciate, and it is this: At least those 1.3 million people had somewhere to run *to*. At least there was a safe and secure mainland to receive them.

Imagine a different scenario. Imagine if all those men, women, and children had not been able to flee at all. Imagine if all the roads out of town had been blocked for some reason and all escape vehicles sabotaged to boot. What if, instead of the few thousand who couldn't or wouldn't flee Katrina, *all* the people of New Orleans and surrounding parishes were left behind. Picture every last schoolteacher and grandmother and checkout girl and auto mechanic and kindergartner and musician and corporate lawyer all huddled behind those faulty levees as a nuclear-scale storm rapidly approached.

Why imagine this? Because, like the long-ignored warnings about insufficient levees in New Orleans, there are extremely serious warnings out there that Katrina-like disasters could become commonplace along vast stretches of U.S. coastlines in the not-so-distant future. And evacuating inland might not be an option, no matter how bad the storm, because extreme weather events in the heartland (droughts, heat waves, forest fires) will remove the welcome mat. There simply won't be the infrastructure and surplus resources needed to absorb the overflowing humanity.

Ever since Katrina hit, Americans have been asking two fundamental questions: How in the world did this disaster happen? And, could a similar calamity happen where I live? This book will answer both questions in detail.

For starters, Katrina devastated New Orleans because, over the decades, we, as a nation, profoundly altered the basic hydrology of the Mississippi River. The river's massive flood levees directly triggered a geologic chain reaction that obliterated the vast wetlands and coastal barrier islands that once protected the city from hurricanes. By 2005, so much land had disappeared that we had essentially created a watery flight path for Katrina to slam into New Orleans like a plane into the World Trade Center. There was nothing "natural" about this natural disaster. *We* did this.

In March 2003, my book *Bayou Farewell: The Rich Life and Tragic Death of Louisiana's Cajun Coast* was published. It predicted in great detail that a Katrina-like storm would soon destroy New Orleans, leaving thousands of people dead and the national economy bruised. When the hurricane did hit, precisely as foreseen, journalists from around the world began calling me, asking how it felt to be a prophet. How amazing, they said, that I was able to see this disaster coming when so many others didn't.

In truth, I deserve no credit whatsoever for my prediction. Katrina's arrival was as certain as tomorrow's sunrise. There were thousands of pages of reports before the storm, from advocacy groups and government agencies, spelling out the need for better levees and bigger barrier islands to prevent the looming catastrophe. Hindsight is 20/20, and Americans are now outraged by the lack of prior action. Yet the predisaster paper trail was *so* long, stretching to the moon and back, that a journalist like me was just stating the obvious prior to August 2005. Katrina was coming. The facts were as clear as day.

And now something else is coming, something just as obvious but much bigger and even more dangerous. Which leads us to the second question on every American's mind: Can Katrina happen where I live? The answer, unfortunately, is yes, yes, and again yes. If you are one of the 150 million Americans who live within a hundred miles of a coastline—and even if you live much farther inland—you could be inhabiting the next New Orleans. The bad news for you is that there are even *more* studies full of even more scientific data confirming this fact than there were predicting Katrina prior to 2005.

The issue this time is global warming. We are literally altering the sky above us. And be assured, this is not some "junk theory" peddled by Greenpeace extremists. No less a voice than the Bush administration has officially confirmed, on multiple occasions, that global warming is real and is driven by our use of fossil fuels—oil, coal, and natural gas. Worldwide, thanks to climate change, sea level is expected to rise up to three feet during this century and extreme weather events are expected to increase—*according to the Bush administration*.

These two factors—more intense storms and rising ocean levels—mean we are rapidly turning the majority of America's coastal cities into places greatly resembling New Orleans. Thanks to global warming, mountain glaciers worldwide are vanishing, sending meltwater into oceans that are themselves warming and growing in volume. The resulting sea-level rise—again, up to three feet by 2100—means that vast areas of many U.S. coastal cities will soon fall *below* sea level just like New Orleans, and they will require levees to survive, just like New Orleans.

On top of this, along America's Atlantic and Gulf coasts, hurricanes are becoming much more ferocious. Three major scientific studies in the past year alone reveal that rising sea-surface temperatures linked to global warming are driving an observed trend toward much stronger hurricanes. One study by a noted scientist at the Massachusetts Institute for Technology shows that hurricane wind speeds have *doubled* in the last fifty years. This may account for the following astonishing fact: Among the six most powerful hurricanes ever to form in the Atlantic Basin in the last 150 years, three of them—a full half—happened *in just fifty-two days in 2005*: Katrina, Rita, and Wilma.

In 2003, I declared with complete confidence that Katrina was coming. I argued that below-sea-level New Orleans would soon fall prey to a major hurricane because of *human* actions. Now I beseech readers to trust me when I say Houston and Tampa and New York City and Baltimore and Miami are in equally deep trouble. If you want to know what disasters these cities will be frantically fighting against fifty to seventy-five years from now, just turn on your television. Look at New Orleans today. That's the future.

Yet a full year after Katrina hit, we are still ignoring that storm's biggest lesson. We continue to turn a blind eye to global warming the same way we once ignored the dire pleas for stronger levees in Louisiana. History is repeating itself on the largest scale imaginable. The pages that follow will make clear that all of America—and indeed the whole planet—is now like a low-lying land behind broken and insufficient levees, and the water is coming up fast.

But, thankfully, there is a plan to get us out of this mess just as there was once a viable plan to prevent Katrina's worst impacts. It involves the seemingly unlikely aid of hybrid cars and modern windmills and solarized homes. Clean energy is the solution to global warming, and clean energy is as widely available to us today as the dirt below our feet for filling sandbags. We just have to pitch in and pick up our shovels and get to work—right now.

In the end, the metaphors only go so far. We have but one planet Earth, and it is not just another watery Louisiana parish we can vacate and return to when the danger's gone. Our days of running are simply running out.

Soon we won't have any place to go.

1

The Real Reason Katrina Happened

THERE IS PERHAPS no better way to understand global warming and all its implications than to carefully study the history of New Orleans. This may sound odd, but I make the claim with complete sincerity. To appreciate the alarming physical impacts fast approaching all of our coastal cities, and to appreciate the patterns of human decision-making and denial that are steering us toward an overheated planet, you have to understand what happened in and around New Orleans prior to August 29, 2005. You have to understand, first and foremost, the history of flooding in that city. Most of the key lessons concerning both Katrina and global warming flow in rather unexpected ways from the fact that the Big Easy has always been under the threat of too much water.

From the very start, there was flooding in New Orleans. In 1718, the year French colonists first settled along that crescent-shaped bend of the Mississippi River, heavy rains sent river water pouring into the settlement, flooding crude homes made of cypress, moss, and clay.

So began nearly three centuries of human struggle against the third-longest river on earth. Using simple shovels and pack animals, French engineers constructed rudimentary dikes, no more than two or three feet high, along the river's edge and around the fledgling colonial encampment. And people punched holes in coffins to keep them from floating to the surface when the next deluge came.

And that deluge came soon enough, in 1719, overwhelming the original levees and covering everything with muddy Mississippi River water. The dikes were immediately made taller. More dirt was moved into place. And the ring of protection was expanded as the city grew and prospered on the commerce of rumrunners and slave traders and fur merchants and pirates.

Of course, those early French settlers, beset by mosquitoes and fever and Indian attacks, had no way of knowing their shovels would help set the table, 287 years later, for the worst "natural" disaster in the history of North America. They never intended, with their rudimentary dikes, to start a domino effect that would obliterate entire barrier islands hundreds of miles away and lay waste to a million acres of buffering marshes that made human life along the Louisiana coast possible to begin with.

But in time, those cypress huts, ringed by waist-high dikes, would give way to skyscrapers kept dry by towering levees and flood walls and pumps and spillways and locks and jetties, all designed to corset the lower Mississippi River. The result, by 2005, was a weirdly walled city, situated mostly below sea level, with the northward-moving Gulf of Mexico pounding at the door.

Again, no one intended for this to happen. Not the royal engineers of the king of France, nor the hard-hatted technicians for the U.S. Army Corps of Engineers. But the connection is direct and emphatic, traveling in straight-line fashion from the sound of those first French spades entering the soil to the thump-thump-thump of helicopters rescuing parched survivors from Lower Ninth Ward rooftops in New

Orleans; a direct line from the slaves with wheelbarrows bolstering nineteenth-century levees to their descendants on the I-10 overpass in 2005; from the diesel-powered mechanical shovels of the twentieth century to the drowned bodies floating down Napoleon Avenue in the early twenty-first century.

When Jean Baptiste Le Moyne, sieur de Bienville, landed in 1718 at the site that would become Nouveau Orleans, the coast of Louisiana looked nothing at all like it does today. The land between that original settlement and the Gulf of Mexico featured dark, dense tracts of forests full of live oaks and honey locusts, adorned with Spanish moss and dwarf palmettos and creeping vines hanging from verdant upper canopies.

Moving south, these forests gave way to vast freshwater marshes and swamps, with swaths of green spartina grass mixed with stands of regal bald cypress, water tupelo, and pumpkin ash. Still farther south came a nearly endless expanse of saltwater marsh, stretching on and on and on to the horizon, teeming with herons and ibises and migrant songbirds along meandering bayou streams thick with redfish and mullet rising and splashing through the rich salt air.

Finally, reaching the edge of the Gulf of Mexico, lay a formidable network of tall and broad barrier islands. These islands were a veritable fortress, literally long walls set stubbornly against the powerful Gulf. Some, like Isles Dernieres, were twenty to thirty miles long and would, in time, support vibrant human communities with hotels and homes as well as crowded rookeries for pelicans and terns and roseate spoonbills.

This is what coastal Louisiana looked like on the maps of the early explorers and state cartographers in the 1700s and 1800s. Even as late as 1916, in the junglelike forests southwest of New Orleans, filmmakers made the very first Tarzan movie, documenting the thriving ecological grandeur that characterized most of south Louisiana.

But by 1999, when I first began exploring this very same coast as a journalist, almost everything described above was gone. In the course of a few short generations, the land had largely vanished. Even now, I can scarcely comprehend this fact, much less the cultural and economic calamity that has followed.

I stumbled on this story quite by accident when an editor for the *Washington Post* asked me in the spring of 1999 to do an article for the Sunday travel section set somewhere in Louisiana. Hoping to immerse myself in the local culture, I grabbed a sleeping bag and backpack and literally hitchhiked by boat through the state's coastal bayou region, traveling and working on shrimp trawlers and crab boats and oyster vessels.

In the watery coastal parishes south and southwest of New Orleans, I spent many days and nights with Cajuns who speak French better than they speak English and still make their living off the land by fishing or trapping or alligator hunting—or all three. I met men with nicknames like Papoose and Dirt and Gator preceding last names like Boudreaux and LeBoeuf and Domangue. Many had dropped out of grade school to help their poor, backcountry Catholic families, leaving them, as adults, unable to read or write but with encyclopedic minds nonetheless.

The Cajun boat captains who picked me up, backpack on my shoulders, would sit proudly in their wheelhouses, many of them steering only with their bare feet while perched atop tall wooden stools and strumming and singing "French" songs on the guitar. Then it would be off to expertly repair a net winch or diesel engine in record time before working up an exquisite plate of red beans and rice in the galley. These were men who could construct an entire wooden shrimp boat with no blueprints whatsoever, measuring and cutting and bending boards only by sight.

Into the Gulf these men took me, and into the coastal marshes, and up and down the serpentine bayous of Louisiana, showing off their knowledge of the natural world by, for example, teaching me how to identify underwater fish just by the swells they make on the surface.

Filling my ears on these sojourns was more than just the exotic cry of birds and the low whistle of wind through the marsh grass. There was language. For hours I listened as Cajun fishermen spoke to each other in the lively, archaic version of French that their ancestors brought to North America centuries ago. Having survived a long history of persecution—in France, in Nova Scotia, and in the American colonies—the Cajuns arrived in Louisiana in the 1790s and today over half a million still speak or at least comprehend a version of French preserved like a wasp in amber. It has the sound, to a modern Frenchman, of what the English of Shakespeare's day would be to the average American today.

But mostly these people spoke to me in a Louisiana English peppered with French color. "*Mai tee friar*, don't be scared by dat *tempet*, she's coming." Translation: "Don't worry about that little storm over there, friend."

And there was more. Eavesdropping on the two-way radios standard in most wheelhouses, I would hear the twangy Texas voices of oil workers heading out to Louisiana's vast constellation of offshore oil and gas platforms. I would hear, too, the exotic language of exiled Vietnamese shrimpers who've fished these waters since the 1975 fall of Saigon, drawn to America's own version of the Mekong Delta. And completing the ethnic gumbo flowing over the radio were French-speaking Native Americans, part of the fifteen-thousand-member United Houma Nation. These people, driven by European settlers over the centuries to the farthest ends of the bayou country, survive today as expert fishing people with a culture all their own.

Riding on these various Louisiana boats, I would pinch myself for my Huck Finn good fortune. With my belly bursting from the generously shared meals of red beans and rice and venison sausage, I would take in the tangerine orange sun setting over a lovely expanse of gorgeous gold marsh grass. Overhead, roseate spoonbills, pink as flamingoes, winged toward distant nests while, in the water, dolphins rose and dove around the boats, chasing big schools of speckled trout.

I had long ago assumed that such a place was no longer possible in

America. Having traveled all across the United States and seen the franchised homogeneity seeping into even the farthest, most remote corners of our land, I was convinced it was no longer possible to fall off the map in America and travel through a semiwilderness landscape with people barely recognizable as my countrymen.

But it is still possible along the coast of Louisiana. Or at least it was. It was possible before August and September of 2005. It was possible before Katrina, and then Rita, hit the southern coast of the United States.

That's when that extensive and complex system of levees, a mainstay of life for so many years, finally killed most of coastal Louisiana.

But not in the way you might think.

OF ALL THE AMAZING THINGS I experienced during my hitchhiking tour through the Louisiana bayous—the food, the music, the language, the wildlife—the thing that impressed me most was none of the above. It was the flooded cemetery. There in tiny Leeville, Louisiana, along Bayou Lafourche fifty miles southwest of New Orleans, was an entire cemetery of tombs and headstones standing in open water.

Equally strange, farther up this same bayou, was an entire stand of oak trees—a whole forest—sitting half-submerged and lifeless in open water. The trunks were eerily sun-bleached and leafless, looking like unburied bones. And nearby, yet another weird site: telephone poles all along the shoulder of a two-lane bayou road, standing in water that lapped the very edge of the asphalt.

My host at that moment was Papoose Ledet, a forty-five-year-old Cajun shrimper from Golden Meadow, Louisiana, who worked his trawler with his two teenage sons. When I asked him about the submerged cemetery and the dead forest, he answered immediately in his heavy Louisiana accent.

"All the land," he said, "it's sinking. It's sinking real bad."

"Sinking?" I responded. "Why in the world is the land sinking?"

So began my education in what, in my mind, is still one of the greatest environmental calamities in America today and a stark foreshadowing of what global warming will bring to the rest of our coastal regions. Sadly, it took working-class Louisiana fishermen, many of them classroom dropouts, to school me in the root causes of the national emergency we now call Katrina. I had traveled the world, been to graduate school, and I considered myself a devoted conservationist. But like virtually all Americans living outside of south Louisiana, I knew nothing about the secret horrors unfolding in the bayous.

Papoose Ledet, steering his boat with his bare foot, patiently guided me though the basic geology of the lower Mississippi River. At the mouth of every great river system in the world, he said, there are two basic geologic phenomena at work. The first one, the one most people are familiar with, is flooding.

The whole ragged sole of the Louisiana boot was created by seven thousand years of natural river flow and flooding from the Mississippi. Draining nearly two-thirds of the lower forty-eight states, from the Montana Rockies to the Appalachians of New York state and everything in between, the river annually transported to Louisiana unfathomable quantities of sediments and nutrients. Each spring, heavy rains across America would cause the river to jump its banks near its mouth and deposit that liquid soil across the bayou coast, thus building the marshes and the barrier islands, building the very soil below the streets of New Orleans, sculpting the entire wetland coast of Louisiana.

This part of river delta dynamics is fairly well known. What is less well known is the second major geologic phenomenon at work in a delta system: land subsidence. The entire coastal landscape created by a river naturally subsides, or sinks. That's because the deposited alluvial soil is very fine and silty and inherently unstable. Over time, its water content is squeezed out and the embedded organic material decomposes. The soil literally shrinks in volume as a result. It compacts upon itself.

Coastal Louisiana is no exception. The entire land platform of south Louisiana sinks naturally. It sinks a lot. Year after year.

Historically, what has counterbalanced this sinking has been the deposit of new sediments from new flooding, so much so that net land *building* occurred despite the subsidence.

Again, this process, continuing for seven thousand years along the Gulf coast, created the land below New Orleans and the surrounding region. Then, in 1718, the French arrived. Their original flood levees, continuously maintained and expanded to the present moment, have completely erased half of the ancient, natural equation. The lower Mississippi has been completely tamed. It no longer floods. Those great earthen levees, as tall as three-story buildings in places, today channel the river straight out to the open water of the Gulf of Mexico. Tens of millions of tons of sediments each year, flowing down from two-thirds of America, now spill right over the edge of the continental shelf and into an oblivion of deep seawater.

So today, owing to human interference, the time-honored creation of land at the great mouth of the mighty Mississippi River has come to a screeching halt.

But not the subsidence. It continues unabated.

Every second, every minute, every hour, all of south Louisiana sinks and sinks and sinks. The land platform subsided approximately two feet during the twentieth century alone, and it shows no sign of slowing down.

THIS IS WHY, in the spring of 1999, riding on Papoose Ledet's trawler, I saw drowned forests and row after row of cemetery tombs in the water. It's why every day in Louisiana, *even without hurricanes*, fifty acres of land turn to water. It's why every ten months, an area of land the size of Manhattan joins the Gulf of Mexico along this Cajun coast. It's why, just since World War II, a land mass equal to Rhode Island has been subtracted from south Louisiana. Down and down and down the land sinks in a process as natural as rainfall and summer heat but, minus replenishment from new flooding, as destructive as a wrecking ball to all life along the coast.

This unforgiving phenomenon—subsidence—also finally explains the very weird and perplexing fact that most of the city of New Orleans is below sea level. In the weeks after Katrina I was utterly amazed at how the media completely and uncritically accepted the "New Orleans is a giant fishbowl" phenomenon. Virtually no one stopped to ask why the city was below sea level. Tampa Bay, another coastal city, is not below sea level. Nor is New York or Houston or San Diego. Why New Orleans?

The answer is subsidence. When the French arrived three hundred years ago, the entire area was at or above sea level. But since then, the Big Easy—the skyscrapers, the Superdome, the French Quarter, all of it—has been sinking right along with the rest of the coast, year after year, inch by inch.

This, ultimately, is why Katrina happened. The hurricane became the biggest storm disaster in U.S. history not because the hurricane levees *failed*, but because for centuries the river levees *held*. The levees left the land starved for new sediment while the natural process of subsidence never paused. This is what turned New Orleans into the famous fishbowl. It's what created the giant depression, surrounded by earthen walls, that was ready to be catastrophically filled with water by 2005.

But that's not all. While lowering and transforming the city from the inside, these same levees were destroying the buffering landmass between the city and the coast, effectively turning two million Americans into a magnet for whatever terrors the Gulf had ready to throw at the shore.

Overdevelopment along U.S. shorelines has been a problem for years, with people voluntarily putting themselves in harm's way, moving in greater and greater numbers to attractive ocean shores and, unfortunately, the inevitable storms. Think Florida panhandle. Think Outer Banks of North Carolina. All of those people moved closer to the ocean. But in Louisiana, the opposite is true. Thanks to subsidence and erosion, we brought the ocean closer and closer to the *people*.

So I was badly mistaken during those early hitchhiking trips through the Louisiana bayous. Riding atop trawlers and crab boats, taking in the broad landscape of salt marsh and the profusion of avian and aquatic life all around, I naturally assumed I was observing a world of abundance. But what I was seeing was just a thin relic of what once was.

The telephone poles immersed in water told the truth. So did the lighthouses, far out in the Gulf, alone in open water, when before they had been on solid land. So did coastal towns like Houma, Louisiana, where the city fathers struggled to keep their drinking water from turning salty. Cartography also told the story. The historic maps of Louisiana from 1750 to 1850 to 1950, followed by satellite photographs from the late seventies forward, show a coast that was once full and muscular and unbroken giving way by 2005 to something wispy and tattered and full of gaping holes.

A million acres of wetlands had disappeared and the once fortresslike barrier islands were almost completely gone. Isles Dernieres had, by the year 2000, fragmented into four separate islands, widely spaced apart, and had lost approximately 80 percent of its previous landmass.

But as if this picture of subsidence and destruction were not enough, the U.S. fossil fuel industry made things even worse. To gain access to the vast deposits of oil and natural gas spread all along the Louisiana coast, companies like Texaco and Amoco dredged ten thousand *miles* of canals and shipping channels through the marsh over the last seventy-five years, triggering massive erosion.

So at the same time that the entire land platform of south Louisiana subsides at a rate of about two feet per century, the terrain itself constantly erodes from the inside like a moth-eaten cloth. The combination of these two implacable forces makes coastal Louisiana the fastest disappearing landmass on earth, bar none.

Which is why Papoose Ledet, steering his trawler with his bare foot as the smell of red beans and rice wafted up from the galley, could turn to me along the bayous and say with complete assurance: "There's no future for people here. We're all gonna leave this place real soon. Anybody who can read a map can see the truth as plain as day. The Gulf's gettin' closer and closer. Sure as I'm sitting here, we're gonna get wiped out."

By "wiped out," of course, he meant a specific, catastrophic event was coming. It was simply a matter of time. Only luck had kept it from happening already.

WHEN THOSE EARLY FRENCH SETTLERS, dressed in homespun clothing and straw hats, first began building crude levees along the Mississippi River, they did so for perfectly rational reasons. They didn't want their children to drown. They didn't want their homes inundated and their crops destroyed. It made complete sense to build levees. As long as their community stayed dry, the French realized, a world of riches "inscribed by the hand of God" would come to them. These fruits were assured by the settlement's obvious location as a strategic port city and by the limitless natural bounty all around.

But staying dry, as we now know, would bring results never envisioned by Jean Baptiste Le Moyne. A vexing undercurrent of human activity worldwide, something I like to call the Law of Unintended Consequences, would doom the coast and its modern inhabitants. This law, which can also be expressed as the "Sorry, we didn't mean to" postulate, grows from a basic truth about our world that we seem unprepared to grasp: Everything in nature is connected to everything. Nothing in an ecosystem, big or small, happens in isolation. Or, as

Sierra Club founder John Muir observed nearly a century ago: "When we try to pick out anything by itself, we find it hitched to everything else in the Universe."

From there it's pretty straightforward: If you profoundly disrupt one major aspect of a natural system as colossal as the lower Mississippi River, you profoundly disrupt all aspects of that system. If you build towering flood levees along the river, the Law of Unintended Consequences dictates that you subtract a Manhattan-size area of land from the coast every ten months and you invite the Gulf of Mexico well into the interior of America.

But framing the issue in terms of natural systems and invoking the name of John Muir gives the impression that this is just another "environmental" story. It is not. That's because all the great riches first envisioned by those early French colonists eventually arrived in full. Indeed, by 2005, no other region in the country could claim so many critical roles in American daily life as south Louisiana.

The ports of Louisiana, including New Orleans, represent the country's doorway to the globe, handling more than a fifth of American exports and imports every day. This includes nearly three-quarters of all grains heading to international markets from Midwestern farms. The industrial and agricultural might of America quite literally meets the world through the waterways of Louisiana.

The region is also the hub of "America's Energy Coast," producing more oil and natural gas from four thousand offshore wells than we import each year from the kingdom of Saudi Arabia. Refineries, chemical processing plants, and the nation's Stategic Petroleum Reserve round out a massive energy infrastructure that Americans would soon appreciate each time they reached for the gasoline pump or switched on the thermostat for winter heat.

But oil is not the only thing lured from these waters. Just ask the people of Morgan City, Louisiana, who each year hold their charmingly eccentric "Shrimp and Petroleum Festival." No less than a third of America's domestic seafood comes from coastal Louisiana. Ten million pounds of oysters. One hundred million pounds of shrimp. So many crabs that even along the distant shores of the Chesapeake Bay, a huge share of those famed crab cakes get their basic ingredient from the Pelican State. And the same Louisiana marshes that serve as nurseries for all of the above also provide critical habitat for millions of birds, including wintering grounds for 70 percent of the nation's migratory waterfowl.

And then there's the culture—jazz, zydeco, gumbo, po'boys, crawfish, Café du Monde, Cajuns, Creoles, Tennessee Williams, Mardi Gras, Jackson Square, shotgun houses, and the second-line parade. Seven million Americans once visited New Orleans each year, generating $5 billion in tourism revenue while the rest of the country profited culturally from one of the last great centers for peculiar tastes and original expression.

Until the storm, that is. None of those first French inhabitants could have imagined, in their wildest dreams, with or without the hand of God, such a bounty of riches along this southern coastline. It was so much, so big, that only an equally extraordinary force, guided by just the right circumstances, could have brought it all to its knees in a matter of hours.

But that's what was about to happen. Dozens of Cajun shrimpers had told me it was coming. The drowned forests said it was coming. The successive satellite photos I had held in my own hands, showing all that green land turning blue—and blue and blue—said it was coming. By 2005 we had created the perfect setting in and around New Orleans.

Katrina was on her way.

2

Ignoring the Warning Signs in Louisiana

L ONG BEFORE KATRINA, everyone in Louisiana had a hurricane story to tell. The names are cursed and revered: Audrey, Ethel, Betsy, Camille, Andrew, Ivan. An oysterman shows you a photo of his roof lying three blocks away atop a friend's car thanks to a "Cat 3" thirty years ago. A coastal restaurant tapes a line just below the cash register marking the high-water mark of a storm twenty years back.

But the darkest hurricane stories always involved hatchets and axes. These were the tales of water rising so fast that folks grabbed the children and an ax and scrambled to the attic, busting a hole in the roof just as the air pocket ran out, everyone then clinging to the topside shingles until rescue or death.

In all my travels through south Louisiana, however, I had never met anyone who had actually experienced an attic escape firsthand. The rooftop stories were always set in the distant past, involving people's older parents or long-dead grandparents. Escapes like this simply no longer happened. Modern levees and early-warning satellites made sure of that.

Which is why, staring at my TV on August 30, 2005, I knew the Big One had finally come to the Louisiana coast. On rooftop after rooftop I saw the hatchets and axes flashing, tearing through attic roofs from the inside. In New Orleans, in St. Bernard Parish, in Plaquemines Parish, whole families in wide-eyed disbelief pulled themselves up through the broken shingles. Helicopters buzzed overhead, meanwhile, lowering Coast Guard crews armed with their own long axes to rescue people unable to bust through on their own.

Something few people alive in Louisiana had ever witnessed before was actually happening.

THE EYE OF KATRINA came ashore near the little town of Buras, Louisiana, sixty miles southeast of New Orleans, at six a.m. on Monday, August 29. Soon there was little left of Buras or the rest of lower Plaquemines Parish except collapsed homes and upside-down schoolhouses. Many of the buildings were speared through by flying telephone poles. Shrimp boats lay wrapped around the broken legs of fallen water towers in roadways crowded with unmoored barges and dead cows.

That's what happens when a "high Category 3" hurricane arrives with 125-mile-per-hour winds, a twenty-foot surge tide, and no barrier islands or marshes to slow it down.

The eye then traveled north before hurtling into Bay St. Louis, Mississippi, and obliterating most of the Mississippi and Alabama coast with an astonishing twenty-nine-foot surge tide. In some places, nothing was left. Even brick houses were blown apart, the bricks scattered and hurled far inland. A floating casino landed on and crushed a Holiday Inn in Biloxi. An offshore oil platform lost its mooring and slammed into a bridge in Mobile. At ground zero in Bay St. Louis, the wind was so ferocious it turned two-by-fours into toothpicks. Veteran disaster workers arrived to find something they'd never seen before: only small, fine splinters covering the ground, several feet deep in places.

More than two hundred Mississippians died the day the storm hit, and Biloxi mayor A. J. Holloway said famously, "This is our tsunami."

And then there was New Orleans and the horrifying images now seared into our national consciousness with September 11 ferocity: the Superdome and I-10 overpass, the convention center and 17th Street Canal levee, hungry looters and starving pets and cadavers left on sidewalks or wrapped in sheets and stored in hospital chapels.

What started with sustained winds tearing off roofs and uprooting trees ended with a city drowning. Katrina's enormous surge tide washed over Buras and Plaquemines Parish and then raced toward the city through Breton Sound and Lake Borgne. There was almost nothing to slow the charge. The natural speed bumps just weren't there. Just since the 1930s, perhaps half the landmass between New Orleans and the Gulf had disappeared due to subsidence and erosion.

Compounding matters was a kind of man-made landing strip. Among the ten thousand miles of dredged waterways in Louisiana was something called the Mississippi River Gulf Outlet, or "Mr. Go." Displacing more dirt than the Panama Canal when it was built and destroying twenty thousand acres of wetlands, this seventy-six-mile-long shipping channel was completed by the U.S. Army Corps of Engineers in 1965 as a shortcut from the Gulf of Mexico to the Port of New Orleans for deep-draft vessels. That boat captains largely ignored the channel after it was built wasn't nearly as tragic as the outlet's long-recognized role as a "hurricane highway."

Katrina's surge entered Mr. Go as if entering a huge funnel. The "squeezed" water instantly rose several feet higher and greatly accelerated its approach speed. Then, as if riding a rocket through space, the water flowed right into the heart of New Orleans, quickly "overtopping" and destroying the levees along the city's Industrial Canal. These breaches are what annihilated the Lower Ninth Ward and most of St. Bernard Parish with as much as twenty feet of water.

Flooding in the rest of the city came from the north, pouring in from Lake Pontchartrain. As the hurricane passed east of the city, its swirling counterclockwise winds pushed a tall wave of lake water south toward New Orleans. Though the Pontchartrain levees were not overtopped, the water rose so quickly to within a few feet of the levee tops that the downward pressure of the water's weight wrecked the structures from below. Water seeped though weak soils underneath the massive steel-and-concrete walls along the 17th Street Canal and London Avenue Canal, causing the structures to fragment and burst apart.

Then the water poured in. It raced down streets. It gushed into homes. It rose so fast—a foot per hour in places—that people scrambled to second floors, then attics, then rooftops.

By Tuesday night, August 30, the great saucer of New Orleans was full. Billions of gallons of water had flooded 80 percent of the city and had nowhere to go. Pumps were overwhelmed and quit working and the levee system was now a hindrance, holding the water in. Parking meters and street signs vanished. Clusters of trunkless treetops marked city parks. And everywhere people suddenly moved about on anything that floated: air mattresses, laundry baskets, doorless refrigerators with wood planks as paddles. Katrina had turned the city into "Ground Below Zero," one city official said.

At least one baby was born in an attic while elsewhere people died by the hundreds. Bodies floated unclaimed in sewage-filled streets as the city turned into a giant "damp grave."

As the sun set on the last day of August 2005, New Orleans was a panorama of the surreal. Biblical shafts of sunlight broke through clouds onto a city set fantastically in a sea of brown water, where highway overpasses rose from the water and returned to the water amid miles and miles of nothing but rooftop islands and the sickening sheen of oil and chemical spills. Here and there the upper floors of buildings caught on fire, sending plumes of smoke up toward the Coast Guard helicopters that streaked by on triage runs trying to help thousands of people emerging from stifling attics. A musician sitting near the Superdome surveyed all the misery and pulled out his violin. With a sadness as deep as the surrounding water, he played Bach's famous lamentation, Sonata No. 1 in G minor.

FROM DAY ONE, virtually every elected official and journalist in America began interpreting the great calamity of Katrina as a matter of flawed levees and poor evacuation plans. These factors, to be sure, contributed enormously to the appalling destruction and death in New Orleans. But in the end, they were but symptoms of a much larger illness.

Blame was as common as wet carpet after the storm. Mayor Ray Nagin came up short in many ways, especially in maintaining basic security for evacuees at the Superdome and convention center. Governor Kathleen Babineaux Blanco seemed chronically indecisive throughout the crisis, and lacked credibility condemning the federal government for a lack of evacuation buses when more than one hundred local school buses, meant for the same purpose, were underwater in New Orleans, having never been moved.

And the near-criminal incompetence of the Federal Emergency Management Agency (FEMA) was best illustrated by the résumé of director Michael Brown, whose prior experience for the position centered on upholding judging standards in Arabian horse contests. One day Brown was pursuing a breeder for supposedly liposuctioning fat out of the rear end of a gray mare, and the next, owing to his friendship with George Bush, he was the head of a half-billion-dollar agency with twenty-five hundred employees tackling the storm of the century.

Yet none of this counts for much once the lens is properly widened. It is essentially beyond challenge that Katrina would not have destroyed New Orleans had it struck the state of Louisiana as it looked on maps in 1750 or 1850 or even 1950. Thirty-six years earlier, Hurricane Camille struck the Louisiana coast with winds *stronger* than Katrina, but the storm surge did not reach many of the areas destroyed by Katrina. What happened in the intervening years was, of course, catastropic land loss triggered by subsidence and canal-building.

The loss of wetlands alone essentially doomed the city. Thousands and thousands of acres of marsh grass once blanketed everything south and east of New Orleans. When Teddy Roosevelt was president, Barataria Bay was more grass than water and the marshy peninsula of Plaquemines Parish was three times wider than it is now. And in Breton Sound, the combination of grasses and barrier islands created a muscular bulwark of land. Katrina simply wouldn't have recognized such a landscape. In 2005, she instead skipped over mostly open water on her flanking move toward New Orleans.

Coastal marshes, it turns out, provide more than just critical habitat for birds and shrimp and crabs. Every 2.7 miles of marsh absorbs a foot of a hurricane's storm surge. The friction of those trillions of blades of grass literally slows everything down, dispersing the energy of the onrushing water, stealing much of the hurricane's punch.

Depending on how you measure it, the area south and east of New Orleans has lost at *least* fifteen to twenty-five linear miles of wetlands in the last hundred years. By my own calculations, this means that the Louisiana of Teddy Roosevelt's day would have knocked down at least five to nine feet of the eighteen-foot surge that eventually overtopped the Industrial Canal levee and destroyed the Lower Ninth Ward. And though harder to calculate, a similar land buffer would have significantly reduced the power and speed of Katrina before she pushed water from Lake Pontchartrain up against the weak 17th Street Canal levee.

This sheltering role of coastal vegetation was dramatically illustrated during the 2004 South Asian tsunami. Researchers found that shorelines lined with mangrove forests suffered significantly less damage than areas where the tidal wave met land denuded by human activity. This scientific analysis conducted soon after the disaster suggests that just 30 trees per 120 square yards in a 100-yard-wide belt could diminish the maximum tsunami impact by more than 90 *percent*.

This is critically relevant because when Biloxi mayor A. J. Holloway said "This is our tsunami," he wasn't speaking metaphorically. That's what a hurricane surge tide is: a wide, domelike wave of water. And when you're the state of Lousiana, losing fifty acres of coastal vegetation per day and an area the size of Manhattan every ten months, the tsunamis just get worse and worse with time. And for the rest of the planet, as sea-level rise from global warming wipes out more and more wetlands and other protective coastal vegetation, similar impacts will occur everywhere, as we'll see in later chapters.

Barely three weeks after Katrina, with a cruelty impossible to imagine, the tidal wave of Hurricane Rita crashed into Louisiana. At one point out at sea, Rita was the most powerful hurricane ever recorded in the Gulf of Mexico, with sustained Category 5 winds of 175 miles per hour and a circumference nearly as big as the Gulf itself. She came ashore along Louisiana's western coast as a Category 3, with a ten-foot surge tide that left Hiroshima-like scenes in Cameron, Holly Beach, Erath, and elsewhere. When the water finally receded, caskets were hanging in trees, shrimp boats were in sugarcane fields fifteen miles inland, and forty-one of sixty-four Louisiana parishes were without electricity. Rita's impact was so broad that the surge tide re-flooded New Orleans, two hundred miles to the east, and toppled levees throughout Louisiana's central coast, completely destroying more than ten thousand homes in Terrebonne Parish alone.

All told, at least five thousand Houma Indians and thousands of Vietnamese Americans and hundreds of thousands of Cajuns—people I wrote about in *Bayou Farewell*—were more or less wiped out by the twin hurricanes. Louisiana's oyster beds were under two feet of silt. Louisiana's shrimp boat fleet no longer existed. Louisiana's crab industry was a mess of missing cages and lost floats and sunken boats.

After a century of near-miss "Big Ones" along a shore of rapidly dissolving land, the worst-case scenario had arrived in full. Katrina destroyed the Louisiana coast from the Mississippi state line to Grand Isle, and Rita pretty much finished the job from Grand Isle to the Texas border.

◆   ◆   ◆

TWO WEEKS AFTER KATRINA'S LANDFALL and the flooding of New Orleans, conservative columnist George Will wrote, "It is likely that Katrina's lingering reverberations will alter the makeup of the nation's mind far more than 9/11 did."

The sheer scale of the catastrophe seemed to bolster that claim. When critics faulted New Orleans mayor Ray Nagin for supposedly falling short of the leadership standard set by Rudy Giuliani during the 9/11 attacks, Nagin pointed out that, as awful as the World Trade Center attack was, it destroyed less than ten square blocks of New York City. Katrina harmed *every* square block of New Orleans and directly affected a surrounding area the size of Great Britain. Comparisons were essentially absurd.

The nation's mind turned to many issues after the hurricane, not the least of which was the shame of so many impoverished black Americans crowded atop the overpasses and inside the Superdome, bearing a fantastically disproportionate share of the suffering. And Americans suddenly wondered why we were spending so many billions rebuilding schools and sewage-treatment plants in Iraq when we desperately needed to rebuild them along our own Gulf coast.

One storm hit America, and in a matter of hours exposed a dense tangle of previously hidden fault lines on race, national security, public health, the economy, and the environment. But energy was the real Achilles' heel highlighted by the storm, bringing perhaps the most pain and panic to the largest number of Americans.

Nearly one thousand manned oil rigs and platforms were evacuated as Katrina approached in the Gulf, and the storm completely destroyed forty-six of those structures. One platform, originally located twelve miles off the Louisiana coast, washed up onshore at Dauphin Island, Alabama. And more than half of all U.S. refineries along the Gulf were put out of action, some for an entire month.

At its peak, 95 percent of all Gulf oil production and 85 percent of all natural gas production was shut down, representing about 1.4 million barrels of oil per day and 8 billion cubic feet of natural gas. The oil impact alone was equal to what was lost per day to the United States when Saddam Hussein invaded Kuwait in 1990.

As a result, gas station owners across the country reported shortages, and prices rose above $5 per gallon in some regions. A government hotline logged more than five thousand complaints of alleged price gouging in September. But even as gas prices eventually eased thanks to imports from Europe and the opening of the Strategic Oil Reserve, natural gas prices stayed extremely high. Long-term rising demand collided with the disruptions of Katrina and then Rita to raise home-heating costs a whopping 40 percent across the nation for the winter of 2006. Suddenly, a lot of Americans knew where their home-heating fuel came from.

And in the Gulf states, jobs of every conceivable sort simply vanished; nearly half a million gone overnight. In New Orleans alone, eighty thousand tourism jobs were erased with a $5 billion annual impact on the economy.

Meanwhile, the Port of New Orleans remained 80 percent closed a full month after Katrina. Hundreds of port workers were scattered to other states and critical rail and highway links to the port had been torn apart. A logjam of agricultural exports ensued and imports of everything from coffee to concrete were expensively rerouted to ports as far away as Florida and Texas.

This cumulative damage, which included the annihilation of a billion-dollar seafood industry and big projected jumps in property insurance from coast to coast, dampened economic growth for the nation as a whole at the close of 2005.

As for the people of hardest-hit Louisiana, they had fled mostly north into the nation's interior, moving upward as if into a giant metaphorical attic and then out onto a rooftop. And that's where things stood for months after the storms. As a culture, as a society, coastal Louisiana was on a battered roof, hatchets and axes tossed aside, still alive, everyone hungry and thirsty and tired and scared, surrounded by still-perilous waters, waiting for the rest of the country to lend a hand and help fix what should never have happened in the first place.

THE VERY IDEA that suicide terrorists could commandeer commercial airliners—the planes bloated with fuel meant for West Coast destinations—and then fly them into the World Trade Center was simply unimaginable to everyone in America. The evil was so novel, the tactic a pure masterstroke, that no one could have predicted it.

The calamity of Katrina, on the other hand, was probably the most widely predicted "natural" disaster in human history. Anyone with even passing knowledge of the situation prior to August 29, 2005, knew that Katrina was coming. It was reported by journalists, described by politicians, discussed at academic conferences, simulated on computer models, outlined in government reports, and routinely predicted by every last Louisiana fisherman I ever met.

On June 1, 2005, just thirteen weeks before the storm, U.S. senator Mary Landrieu of Louisiana escorted twenty-five schoolchildren to the French Quarter, everyone dressed in bright orange life jackets. The kids then gathered on lovely wrought-iron balconies, a full story above the street, as a giant blue tarp was draped below them to illustrate just how high the water would reach if the right hurricane hit. Exactly ninety days later, the water reached that mark throughout much of New Orleans—not blue, but brown with sewage, chemicals, and gas.

That the city's hurricane levees were deeply flawed was near the top of Louisiana's list of worst-kept secrets. For many years the U.S. Army Corps of Engineers had warned that New Orleans could not withstand anything more than a relatively weak Category 3 hurricane. In 1995, when a ferocious rainstorm led to six deaths in the city, the Corps petitioned Congress for the $430 million already authorized to improve levees and pumping stations. But only a small fraction of that money was ever appropriated.

In 2004, the New Orleans *Times-Picayune* reported that the Bush administration was investing less than 20 percent of what was necessary to adequately fortify the city's levees, as determined by the Corps of Engineers. And that same year, the Bush administration slashed funding to the Corps' New Orleans district by more than 80 percent.

All of this came after the Federal Emergency Management Agency, in a 2001 report, declared that a hurricane blow to New Orleans was one of the three most likely megadisasters to occur in America. And barely a year before that prediction came true, a gathering of state, local, and federal officials used computer models to examine the impacts of a fictional "Hurricane Pam." That exercise showed city levees collapsing from a surge tide arriving almost unimpeded from the Gulf. It showed "the bowl" of New Orleans catastrophically filling with water. It showed a million people evacuating the region and half a million buildings destroyed. It implied thousands of fatalities from drowning and subsequent hardships.

And yet, on September 1, 2005, three days after Katrina hit, President Bush stood before reporters and said, "I don't think anyone anticipated the breach in the levees."

The second most poorly kept secret in Louisiana was the fact that the entire coastal land mass was in tatters. Major articles describing the crisis had appeared in the *Baton Rouge Advocate* (1999), *Scientific American* (2001), New Orleans *Times-Picayune* (2002), the *Washington Post* (2003), and *National Geographic* (2004).

The most detailed account of both the causes and solutions associated with land loss came with the publication of my book *Bayou Farewell* in March 2003. In the opening pages I wrote: "A devastating chain reaction has resulted from the taming of the Mississippi [River], and now the entire coast is dissolving at breakneck speed . . . and New Orleans itself is at great risk of vanishing. A major hurricane approaching from the right direction could cause tens of thousands of deaths."

Mike Foster, a conservative Republican, was governor of Louisiana when *Bayou Farewell* was published. An avid hunter and fisherman who was also concerned about impacts on big business, he publicly committed himself to waging "jihad" against the problem of coastal erosion, pushing the issue to the top of his legislative agenda.

And thankfully a solution was squarely within reach. As far back as the early 1970s, geologists at Louisiana State University had established that a series of feasible engineering projects not only could counteract much of the subsidence but could actually create *new* barrier islands and wetlands fairly rapidly. This would involve permanently closing some of the most damaging canals and channels, like the Mississippi River Gulf Outlet. It would also involve harnessing the great land-building power of the Mississippi River itself. By building several gatelike "control structures" right into the levees of the river, the sediment-rich water could be released and then surgically guided via pipelines and man-made canals to areas in greatest need of wetland and barrier island development.

Not a single politician or credible scientist challenged the ability of this plan, formally known as the Coast 2050 plan, to restore Louisiana's shores. The problem, before Katrina hit, was simply the price tag: $14 billion. Though equal to just six weeks of spending in the Iraq war or the cost of the "Big Dig" tunnel project in Boston, it was still far more than Louisiana could afford on its own. Serious federal help was needed.

Which is why, when *Bayou Farewell* was published, Governor Mike Foster purchased fifteen hundred copies of the book and gave one to every member of the Louisiana legislature and every member of the U.S. Congress. As an added step in his "holy war" to restore the coast and stave off disaster, he personally sent a copy to President Bush. The plea could not have been delivered in a more urgent, direct, and personal way straight to the White House.

But Foster's requests for full funding of the Coast 2050 plan were ignored by the president. So were multiple pleas from Louisiana's bipartisan delegation to Congress. In January 2004, soon after being elected, Louisiana's new Democratic governor Kathleen Blanco met personally with Bush and renewed the urgent call. But again nothing.

Despite the growing crisis, the president somehow found millions of dollars for the restoration of wetlands in Iraq while Louisiana lay dying on the table. Saddam Hussein had drained a vast area of wetlands inhabited by some of his staunchest enemies, the so-called Marsh Arabs, a fiercely independent people living near the mouth of the Tigris and Euphrates Rivers. Following the rule that the enemy of my enemy is my friend, the Bush administration in 2004 proposed spending *ten times* more federal money restoring wetlands for these Arab people than for restoring the Louisiana coast for Americans. Congress ultimately denied the request, but Bush actually sent wetland experts from Louisiana to Iraq anyway to help spend the millions of dollars he persuaded Japan and Italy to invest in the project.

Only in the summer of 2005, just before Katrina struck, did Congress finally appropriate $570 million in new coastal restoration money for Louisiana. But the sum was fantastically short of the $14 billion needed and was itself spread out over four years. Even this amount was allocated over the objections of the White House.

Meanwhile, as all the foot-dragging and Iraq war distractions went on, the Katrina disaster drew closer and closer and closer, like a train barreling down the track. Again, unlike 9/11, we saw this event coming from miles and miles away. Notwithstanding the warnings of journalists and engineers and Louisiana governors from both parties, we just stood there on the track year after year as the whistle grew louder and the crossing guard bells clanged faster and the trestles began to shake. We never stepped off the tracks. We never got out of the way.

Why in the world not? What went wrong? What could possibly explain such insanity?

3

Why Do Societies Commit Suicide?

THE DEMISE OF NEW ORLEANS and much of south Louisiana can only be described, in the end, as a form of group suicide. The warnings were clarion clear. The preparations nearly nonexistent. The final blow came with an air of deliberation, wrought with knowing.

And while it may be of little solace to the destroyed lives of over one million affected Louisianians, it's important to know that this is not the first time a human society has consciously offed itself. History is littered with such cases, in fact, from the ancient Maya of Central America to the Greenland Vikings to the Polynesian society of Easter Island in the South Pacific.

Pulitzer Prize–winning author and geographer Jared Diamond describes in detail these and other "group implosions" in his 2005 book *Collapse: How Societies Choose to Fail or Succeed*. Sifting through the wreckage of past mistakes, Diamond identifies five major "interacting" factors that have brought societies low: hostile enemies, climate change, self-inflicted environmental degradation, adverse changes in trading partners, and, finally, a society's political, economic, and social responses to the aforementioned factors.

It might strike some as inappropriate, even ridiculous, to compare the collapse of a modern U.S. coastal area with "primitive" long-ago societies. But the opposite seems true to me. If a modern society—with modern communications and resources and education—can kill itself, there must be some deep-seated, ancient tendency at work here.

The demise of the highly complex Polynesian culture on Easter Island five hundred years ago stands out as a particularly useful case study when considering the catastrophe of New Orleans and south Louisiana. Neither society was at war when calamity struck. Nor were disruptions in basic commerce and trade at fault.

Instead, both cultures happened to develop on ecologically fragile landscapes. Both cultures made seemingly rational choices to exploit and dominate that landscape. And then, as these landscapes began to unravel with the full knowledge of decision makers and average people, neither society acted in time to ward off cataclysm.

When Polynesian explorers first arrived on tiny, isolated Easter Island around A.D. 400, they found a subtropical paradise of forests, animals, and fertile volcanic soil. Enormous palm trees eighty feet tall and six feet in diameter towered over a lush layer of woody bushes, shrubs, herbs, ferns, grasses, and fantastic tree daisies. Among the diverse wildlife were countless nesting seabirds ready for the stew pot: petrels, albatrosses, fulmars, terns, frigate birds, and prions. This "environmental abundance" eventually sustained a society of up to thirty thousand people so advanced they produced hundreds of enormous stone statues that centuries later still inspire awe worldwide.

But by 1722, when the first European explorers arrived, the island was a lunar landscape of destroyed soils and treeless contours. Dozens of native trees and every last native land bird had become extinct on the island. The native animals included nothing larger than insects. There were no bats, lizards, or even snails on the island. And the human population had plummeted from thirty thousand to just two thousand souls lacking enough wood even for cooking fires. They burned mostly grasses to prepare a menu that included each other. They had become cannibals.

What happened? The Easter Islanders cut down every last tree on their island, that's what happened. One by one the trees fell until there were no more, and so the people starved. The giant palm trees had provided edible nuts and oil. More important, from the enormous palm trunks the islanders had carved giant sea-faring canoes. These canoes were used to harpoon the staple of their diet: dolphins.

But when the trees disappeared so did the canoes and, by extension, the critical dolphin meat. Deforestation also triggered catastrophic soil erosion, making productive agriculture impossible. Island statuettes from the later years tell the story, showing people with sunken cheeks and exposed ribs.

"I have often asked myself, 'What did the Easter Islander who cut down the last palm tree say while he was doing it?' " Diamond writes. He goes on to list a number of factors that seem to explain the "failures of group decision-making" that lead societies like Easter Island to slit their wrists and then stand by as every last drop of blood pours to the ground.

One big reason involves the pursuit of short-term gains at the expense of long-term survival, he writes. It was just easier *right now* to cut down the trees for canoes to get the dolphins than to devise a more complex yet sustainable livelihood on Easter Island. Another factor, according to Diamond, involves conflicts of interest within a society, when one subgroup prospers by pursuing activities that harm the rest of society.

There is also the problem of elite, unresponsive leaders who insulate themselves from the results of their actions. "Easter Island chiefs made choices that eventually undermined their [society]. They themselves did not begin to feel deprived until they had irreversibly destroyed their landscape," writes Diamond.

These human failings, of course, work hand in hand with the Law of Unintended Consequences. Like the French colonists in Louisiana who had no intention of ultimately destroying the entire coast with their early levees, the first Easter Islanders did not intend to reduce future generations to cannibalism when they discovered that the island's massive palm trees made for excellent canoes. But everything is connected to everything. Without the trees, the fragile island soil couldn't survive the erosive effect of wind and rains. Without soil, new trees couldn't grow. With no new trees, there could be no more canoes. Without the canoes, there was no more dolphin meat for humans to eat.

The circumstances in Louisiana were different, of course, but the results were the same. In Louisiana, the clever human tool similar in effect to the Easter Island canoes was the system of river flood levees. The levees made prosperity possible for a very long time, but their continued use ultimately destroyed the entire landscape upon which the society was based.

That Louisianians and the rest of the nation allowed this regional American "collapse" to happen involves factors both complicated and disturbing. Probing them now offers vital lessons in the wake of Katrina and, just as important, sheds direct light on the ultimate environmental and societal collapse now fast approaching us with our full knowledge and with warning bells clanging loudly: global warming.

As was surely the case on Easter Island and in other failed societies, the people of Louisiana did not at first realize there was a problem. Ever since people had begun living along the coast, land had been disappearing naturally only to be restored later by the next big flood or a change in the lower Mississippi's route to the sea. For centuries people had watched wetlands turn to water. It was normal.

What wasn't normal were the new river levees. Once the Army

Corps modernized and essentially perfected its Mississippi River levee system in the early twentieth century, the disappearing land everyone continued to see ceased to be natural. It was now net land loss. There would be no more restorative flooding. The entire coast was vanishing.

But for years even scientists didn't fully understand what was happening. Then, in the late 1960s, a bright young geologist at Louisiana State University began an extensive series of studies. Using historic maps and radiocarbon dating technology and tests of the river's sediment load, Dr. Sherwood "Woody" Gagliano produced scientific data showing conclusively that the seven-thousand-year period of land building along the Louisiana coast had come to a screeching halt, and now the Gulf of Mexico was rushing north toward New Orleans. By 1973, Gagliano also proposed a technologically feasible plan to rebuild coastal land through controlled releases of river water. Gagliano's work contributed enormously to the world's understanding of basic river delta dynamics.

Yet despite the serious and scientifically unchallengeable implications for all life along the Louisiana coast, Gagliano's dramatic first shot across the bow did little to spur protective action in the early 1970s. Many of the reasons dovetail closely with Diamond's analysis of other societies.

First, short-term gain took precedence over everything else. Times were way too good in south Louisiana to worry about long-term planning. The seafood industry was booming in part because all that drowning and decomposing marsh grass actually "composted" the water, stimulating bigger harvests of crabs and shrimp and oysters in a cruel way. The oil and gas industry was booming as well, and tourism had never been better thanks to America's newfound love affair with all things Cajun. With all this, it was easy to look the other way for years as a football field of land turned to water every half hour.

There were also conflicting interests. One powerful sector, the oil and gas industry, was prospering at the expense of everyone else. Among Gagliano's breakthrough findings was the fact that the oil industry's ten thousand miles of canals were triggering enormous region-wide erosion. Indeed, later studies showed that at least 25 percent of all coastal land loss was from industry canals.

Not happy with these findings and fearing it would be heavily blamed and held financially responsible, the oil industry spent much of the 1970s challenging the science of land loss in south Louisiana, and then simply ignored the data as subsequent study teams reinforced Gagliano's work. This posture seriously delayed action on land restoration, thereby injuring society as a whole. The oil industry's relationship to global warming has been frightfully similar, of course, and equally harmful, as we'll soon see.

Compounding matters in Louisiana was the famously laidback culture of the region itself. The very elements needed to forcefully address a societal crisis—grassroots organizing, coalition building, political direct action—clashed head on with the *c'est la vie* hedonism, quirky individualism, and self-reliance that gave the region its charm. It's easy to find ten Cajun shrimpers who'll offer you a bowl of gumbo on the spot. Getting the same ten to agree on what day of the week it is, much less organize politically, is another matter. It's no coincidence that one of the most popular drinks in the French Quarter is the Hurricane. Just take a sip, *cher*, and ride it out. *Laissez les bons temps rouler.*

Of course, one would like to believe that that's why we have elected leaders, to take care of obvious problems on society's behalf. But, echoing another Jared Diamond factor, most Louisiana leaders were themselves insulated from the problem. Their own homes had not been destroyed—yet. Times were good for everyone, and leaders wrapped themselves up in the common illusion of permanent prosperity.

Thus, year after year, decade after decade, the warnings about the disappearing land went unheeded as did explicit warnings about insufficient hurricane levees in and around New Orleans. The truth of the matter is that, even though Louisiana leaders of both parties began loudly discussing the problem of hurricane vulnerability by the early 2000s and begging for federal help, they waited *way* too long to act, allowing the situation to reach appalling dimensions.

Again, a viable solution was on the table all the while: Let the Mississippi flow from her banks and restore the land in her ancient way. It dated back to the Nixon administration, this engineering proposal. It just wasn't chosen. Simple as that.

Which raises the question: Was the destruction of south Louisiana inevitable? Does the historic record from ancient time to the drowning of New Orleans suggest that certain crises of certain dimensions, especially those with ecological underpinnings, inevitably overwhelm societies? Are such suicides unavoidable? Is it a law like gravity?

The answer, riding in on a soft wind of hope, is no. Some societies in the past have pulled back from the brink of collapse. They've chosen to live.

THOUSANDS OF YEARS AGO, before widespread human colonization, the islands that today make up the nation of Japan were dense and deep-green with forest cover from shore to shore. Coastal areas and valley lowlands featured stout hardwoods like Ubame oaks and Japanese birch while mountain slopes were home to a range of conifers including the majestic larch tree.

Early Japanese communities, their populations small, exploited these forests continuously but lightly. They cleared small agricultural plots and fertilized them with leaf-litter from the forest floor. Fuel wood came from fallen and low-lying branches, and drinking water was kept clean and safe by the intact forest itself.

But this pastoral system didn't last, explains Yale University historian Conrad Totman in his book *The Green Archipelago: Forestry in Pre-Industrial Japan*. By A.D. 600, Japan's forests came under pressure from ruling elites demanding more and more timber to create armies, castles, and religious monuments. Matters reached crisis proportions in the 1600s when two major factors converged: Japan's population reached a historic high of 30 million people and, ironically, a long period of peace came with the Tokugawa shogunate. Peace brought prosperity and the rapid growth of cities and the need for wood for homes, furniture, tools, and shrines. By 1670, much of Japan's once-glorious old-growth forests were completely logged. Soil erosion, floods, mudslides, and barren farmland became common. A collapse like Easter Island seemed inevitable.

But it didn't happen. The reality of increasingly harmed Japanese peasants led rulers to launch one of the world's first and most successful reforestation programs. By 1800, in fact, large tree plantations were satisfying much of the timber needs of the imperial families as well as the booming cities. Meanwhile, "silviculture missionaries" spread throughout the countryside sharing increasingly sophisticated, community-based methods for planting and caring for trees. Vast tracts of managed forests steadily returned to the Japanese islands, restoring to society most of the ecosystem services of tree cover while allowing for sustainable timber harvests.

After teetering on the brink of ruin three hundred years ago, Japan today is again a "green archipelago" of mostly unbroken forest land. The country has an astounding 70 percent or so of its land under forest cover—the most of any industrialized nation even while it has the highest population density of the developed world. All of Japan's people and agricultural production occupy just 20 percent of its land! Nearly all the rest is forest.

How this transformation first began and how it has endured for so many years involves many complex factors. But Totman and other scholars believe the centuries-old tradition of intense cooperation among Japanese villagers has played an important role. This cooperation—born of the need to carefully and fairly regulate water for rice production as well as to defend against bandits—was successfully applied to the careful management of forests for the common good from around 1670 forward.

And Japan, it turns out, has not been alone in charting a more sustainable course for itself, avoiding the fate of Easter Island. Jared Diamond writes of similar environmental success stories that stretch from the highlands of Papua New Guinea to the island nation of Tonga to parts of central and northwestern Europe.

If only Louisiana had rebuilt its wetlands and barrier islands the way Japan rebuilt its forests, nearly half a million Louisianians would not still be displaced with the likelihood that most will never return home. Apart from the cultural differences that contributed to the destruction of New Orleans and environs, what makes the American case especially heartbreaking is just how viable the coastal rescue plan really was.

For years, leading engineers from the government and private sector had been saying that, with the right amount of resources and regulatory backing, they could play the role of Mother Nature and build entire barrier islands between New Orleans and the Gulf of Mexico. Better yet, these long and broad and strong walls of land could be constructed in as little as twelve months. Significant wetland creation would take longer, but even these could be created along the coast to the tune of tens of thousands of acres per year.

This was all possible, of course, because the lower Mississippi River is chock-full of soil. Again, draining thirty-one states and parts of two Canadian provinces, every ten years the river dumps enough sediment into the deep water of the Gulf to cover half of Delaware with a foot of dirt. But how do you get this soil out of the river and into the fragile marshes where it's needed without flooding homes and destroying infrastructure?

Easy, said Woody Gagliano in the early 1970s. Build damlike "control structures" right into the river's flood levees and then open the gates to allow the water to flow out when you want it and where you want it, traveling through carefully designed pipelines and canals. The result would be new land. In 1991, the Army Corps built a modest concrete "diversion" project south of New Orleans that served as a test of this restoration approach. Within a few years, satellite photos showed the diversion had created hundreds if not thousands of acres of new marshland.

With rising public and scientific support, a coalition of south Louisiana leaders in the mid-1990s, representing everything from business interests to the faith community, pulled together a master plan called Coast 2050: Toward a Sustainable Coastal Louisiana. The plan called for a dozen sediment diversions to be built along the river above and below New Orleans. Again, the water would be surgically guided via canals or "slurry" pipelines to where barrier islands and wetlands were needed most in the fight against hurricanes. Virtually everyone agreed this approach, combined with steps like closing the Mississippi River Gulf Outlet, would make quick progress toward resecuring the coast.

The only problem was the cost. This comprehensive rescue package would require $14 billion to carry out, and Louisiana, by itself, couldn't afford it. Washington, meanwhile, was of little help, with federal officials under both Clinton and Bush lapsing into sticker shock. Never mind that this was the only permanent cure to the otherwise fatal disease of land loss. Never mind that $14 billion was the exact cost of the "Big Dig" tunnel-building project in downtown Boston or, later, the cost of just six weeks of spending in Iraq. It was too much.

Besides, the Gulf seafood was still pouring in, the offshore oil was still flowing, and the people weren't exactly rioting in the streets. Let's put this off. Let's not invest just yet. Let's keep looking the other way.

Such was the message from Washington. Those of us who *knew* Katrina was near, meanwhile, and did everything we could to promote this specific rescue plan from the 1990s forward, would routinely shake our heads in frustration and say, "Well, maybe it'll take a catastrophic hurricane wiping out New Orleans before we get the national attention and federal funding needed for Coast 2050. But will there be anything left to save by then?"

One final factor, I think, helps explain why nothing was done to prepare for Katrina, and it is this: At times, some primitive part of our brain can't quite distinguish between science and superstition. Before science, when humans dwelled in the darkness of ignorance and mysticism, there was always some quack "prophet" saying the sky was falling: "Burn all your possessions and join me on the mountaintop and you'll be saved when the sky falls."

Those who foolishly joined the prophet on the mountaintop, of course, lost everything and likely died of exposure and hunger and did not pass their genes on to subsequent generations. Those who were cautious and made the seemingly safe bet to stay behind, they lived and sent their traits forward to posterity.

No doubt there were many ignored prophets on Easter Island who said the sky was falling from deforestation. Likewise among the ancient Maya as their rain forest world disappeared. Ditto among the Greenland Vikings as soil erosion and a refusal to adopt the sustainable ways of the native Inuit people led the Europeans to starve.

The real prophets, regularly outnumbered by the false ones, get lumped in with the latter and so, at select moments in history, everyone in a society gets burned. Unfortunately, modern science seems to have done little to change all this. We routinely accept science and its benefits unless the science says the sky is falling, at which point all those genes of the people who didn't burn their goods and follow the prophet to the mountaintop kick in.

There was no superstition running through the science of Dr. Woody Gagliano when he began saying in the early 1970s that catastrophe was coming to Louisiana. There was no dose of mysticism in the message when he declared we could do something about it with relatively straightforward engineering schemes. It's just that, despite the charts and data sets and satellite maps, he sounded so much like that queer untrustworthy voice from our long-ago mountaintop past.

ON AUGUST 29, 2005, it finally happened precisely as scores of prophets said it would. The blue dome above Louisiana shattered and came raining down on the people of the coast. Katrina's surge steamrolled the parishes surrounding New Orleans and poured into the giant bowl itself, creating a region overnight without housing, without an economy, without people.

Thousands of soldiers in camouflage Humvees patrolled the rubble-strewn streets of the Big Easy. Others manned armed checkpoints to the constant din of military helicopters overhead. The chaotic human-wave exodus from the coast led to a housing crisis across the American south and beyond that made the Dust Bowl migration of the Great Depression look like a beach-blanket picnic.

Within weeks the electric utility serving New Orleans was bankrupt, the pro football team was playing home games in Baton Rouge and Texas, the city had laid off thousands of its workers, and nearly every public school was closed. Mayor Nagin estimated that half or more of the city's residents would never return.

But nothing communicated the nightmare-come-to-earth reality more than the fact that Louisianians began taking their own lives. Trapped far from home with no good prospects, or returning to doomed and devastated neighborhoods only to gather family valuables, they hanged themselves and put guns to their heads and overdosed on drugs in record numbers. The suicide rate in Jefferson Parish, which includes part of New Orleans, doubled in the fall of 2005 compared with the year before, according to the *Washington Post*. Calls to one federal suicide-prevention hotline increased ninefold to nine hundred a day immediately after the storm. Surrounded by looting and drownings and children begging for food, two New Orleans police officers killed themselves in the days after the storm, one in his own squad car, the other right in front of other officers after learning his own wife had died.

These were added to the fifteen hundred people killed by the storm itself across the region. For weeks in New Orleans, new bodies emerged in attics, in obscure alleyways. Rescue workers and journalists told of searching rotting corpses for ID only to have an arm or a leg rip and come free in the search for a wallet.

The sky had truly fallen on Louisiana.

And since it had fallen, it was now a given that Louisiana would finally receive the $14 billion it needed to rebuild the wetlands and barrier islands that could make the coast habitable again. After all, what was $14 billion compared to the estimated $200 billion in economic losses and government recovery spending associated with Katrina?

On September 15, with the entire nation focused and ready to help, President Bush seemed to send just the right message in his dramatic, prime-time speech from New Orleans's Jackson Square, saying, "We will do what it takes and stay as long as it takes" to save the coast.

But it was what he didn't say that was unsettling. Not once in his soaring rhetoric about our "powerful American determination" and about the need to fix levees and improve emergency response plans and confront the legacy of racism, not once did the president say the words *wetlands* or *barrier islands*. Indeed, in a total of six speeches spread over eight trips to the Gulf soon after Katrina hit, he never spoke even in passing about the erosion and subsidence that made the hurricane the monster it was.

So it was little wonder that in its post-Katrina emergency spending package sent to Congress in November 2005, the White House totally dismissed the Coast 2050 plan with a shockingly small $250 million proposed authorization instead of the $14 billion needed.

It's impossible to describe the stunned dismay that seized Louisiana officials and conservationists and journalists connected to the issue when the White House plan was revealed. How in the world could this happen? How could this administration, found totally unprepared for the first Katrina, not see the obvious action needed to prevent the next one? After so much suffering and death and destruction, how could Bush not be bothered to put up the equivalent of a tunnel-building project in Boston or six weeks of spending in Iraq?

Whatever the answers, this national disgrace still stood as this book went to press. Tens of billions of federal dollars are being spent to treat the symptoms: broken levees and collapsed bridges and flawed evacuation plans. But almost nothing for the disease. Which means, quite literally, New Orleans is being abandoned by this government and this nation. Abandoned in the same way that all those tired and hungry people were abandoned at the convention center and on the

I-10 overpass in the days after the storm. The scale is just larger now, the time frame permanent.

"Either the president doesn't get it or he just doesn't care," said Mark Davis, director of the Coalition to Restore Coastal Louisiana. "But the results are the same: more disaster and more death for south Louisiana."

Indeed, every day that the president tells business owners and average citizens to return to New Orleans to rebuild, while still not funding the Coast 2050 plan, is a day he's committing an act of mass homicide. He's actively pushing men, women, and children into the path of the next great tsunami. Period.

The problem, I think, is that when George Bush hears "wetlands" he retreats into a blind ideological hatred of all things "environmental." He hears wasted government spending on pretty birds and liberals' attempts to hamstring sacred corporate business interests. Never mind that these same corporate interests in Louisiana—shipping companies, the oil and gas industry, the seafood industry—are now calling for wetland restoration as soon as possible. But a rigidly fixed worldview is precisely what defines an ideologue, and even self-interest is a frequent casualty.

A friend offers the added observation that human beings typically insist on learning life's biggest lessons the hardest way possible. One world war is not enough. We need two to sell Europeans on the attractiveness of peace. So, too, apparently, with Katrina. One is not enough.

Which means the people and culture of south Louisiana will in fact go the way of Easter Island. It's practically assured. What happened along this coast prior to August 29, 2005, can be justly called a regional suicide. Everything since is a federal mass murder.

But perhaps the Bush administration's response shouldn't come as such a surprise. Why should Bush care about New Orleans when he's simultaneously abandoning *every* coastal city in America? Scientific data from his own federal agencies, published throughout his tenure in multiple federal reports, all say the same thing. This administration is rapidly creating along every U.S. coastline the exact same conditions that wiped out New Orleans. Decisions have been made at the highest levels of the federal government that are luring ocean water closer and closer to the streets of New York and Tampa Bay and Houston and San Diego and Savannah and Baltimore and Los Angeles.

These core national policies, pursued over the objections of nearly every other country on earth, are simultaneously creating *warmer* oceans—much warmer oceans—that are the very breeding grounds for violent, monstrous storms like Katrina and Rita.

Rising oceans coupled with more intense storms? It doesn't take long to see where that leads. By failing to heed the warning signs associated with these deeds—warnings like those in the Big Easy before the Big One came—we are about to bring the Big One to over half the population of America.

4

Global Warming: Same Mistakes, Bigger Stage

AT ROUGHLY THE SAME TIME French colonists were settling the frontier of south Louisiana, a British blacksmith with little formal schooling was developing the world's first commercial steam engine in Devonshire, England. Like the French with their levees, Thomas Newcomer had no intention of harming anyone or anything with this new tool. How could he know he was about to set in motion a process that would eventually alter the weather over every square foot of planet Earth.

All Newcomer knew in 1711 was that this one steam engine, powered by burning coal, could do the work of *five hundred horses*. Imagine the wealth that would come if one simply burned a little coal to create a little steam to power a few pistons that could, in turn, get five hundred horses to do whatever you wanted, from pumping water to turning textile looms to pulling heavy wagons. So began the world's great Industrial Revolution, launched by an engine first called, ironically, an "atmospheric" machine.

In time, locomotives and ships and farm machinery and every conceivable industrial gizmo worldwide would be running on coal and steam. British coal use climbed from 2.7 million tons per year in 1700 to 250 million tons by 1900. Then, in the twentieth century, the use of coal switched from creating steam for engines to creating nearly 40 percent of all the world's electricity to power everything from subway cars to iPods. By 2005, the planet was burning 5.8 billion short tons of coal per year to send electrons hurtling down conductive wires virtually everywhere people could be found.

But there was a problem. A big problem. Way back on Christmas Eve, 1894, sitting in his Stockholm study, Swedish chemist Svante Arrhenius first began calculating the unintended consequences of torching so much fossil fuel. When you burn coal, Arrhenius knew, one by-product was an odorless, invisible gas called carbon dioxide. CO_2 migrates to the atmosphere upon combustion and stays up there for a century or so. It also traps heat. Light from the sun passes easily through the CO_2-laden atmosphere, but infrared radiation, created when light strikes the earth's surface, cannot pass as easily back out to space. If enough people burned enough coal, Arrhenius realized, the entire planet would grow much warmer.

But given the rate of energy use at the time, Arrhenius believed significant warming would take centuries to accomplish. What he didn't know was that a group of American entrepreneurs, working feverishly atop an East Texas salt dome, would soon accelerate that process enormously. On January 10, 1901, after months of physical struggle and investment uncertainty, these Texas "wildcat" drillers managed to drive a drill bit 1,020 feet into the earth only to watch it come right back up in an explosive gusher of oil. What had previously been a relatively scarce liquid fuel from small wells in Pennsylvania was about to be available, thanks to deep-drilling technology, all across the world to the tune of tens of millions of barrels per day.

The petroleum era had begun in earnest, and soon Henry Ford's assembly-line process would mass-produce the cheap cars to match the cheap gas. Unprecedented mobility joined with reliable electricity from coal to produce unfathomable economic growth. By the year 2000, the goods and services produced worldwide every twelve months were equal to *everything* produced in the entire nineteenth century. To keep up, global energy use roughly doubled every thirty years during much of the twentieth century, and carbon dioxide rushed up to the atmosphere like a great river.

Today, climatologists like to call it the hockey stick. Imagine a hockey stick held parallel to the ground with the blade turned up. If you plot the history of atmospheric carbon dioxide levels on a graph, you find that for the past 10,000 years the levels have been very stable, ranging between 270 and 290 parts per million (ppm). That's the long flat part of the hockey stick, the handle.

But beginning around 1750, with widespread use of coal for steam engines, and accelerating dramatically in the twentieth century, thanks to oil, carbon dioxide has risen to the present level of 380 ppm with the prospect of rocketing to as much as 700 ppm or more by 2100. That's the "blade" portion of the hockey stick. It's pointing straight up from the flat line of our past. The heat-trapping power of our atmosphere is growing exponentially greater at this very moment.

But just like Easter Island and Viking Greenland and coastal Louisiana, it has taken a dangerously long time for people worldwide to realize what was happening to the planet; time to comprehend the ultimate lesson in how everything is connected to everything; that if you dramatically alter the chemistry of the global atmosphere you dramatically change everything on earth.

Or, using a baseball metaphor favored by a Louisiana friend of mine, nature always bats last. We humans might get a few singles and doubles and score a few runs for a while, but when we're done and nature steps to the plate, she gets whatever hits and runs are her due and then the game's over.

Period.

* * *

By 1979, enough scientists had detected enough warming and arctic ice-melt and rising sea levels that an international conference was held in Geneva, Switzerland, to discuss the matter. But human-induced climate change was still a "theory" that had earned few converts since Svante Arrhenius's day. Scientists and policy makers agreed in Geneva to continue to monitor the situation.

The warming, meanwhile, didn't pause. Virtually every year in the 1980s was warmer than anything science had recorded prior to 1980. The timing of seasons began to change. Insects hatched earlier. Animals changed their migration patterns. Plants flowered sooner. And extreme weather events began to grow.

Then 1988 happened. It brought one of the hottest, driest summers ever seen in America up to that point. Forest fires burned millions of acres in the western United States and food production fell all across the nation. Midwestern states grew so dry the sky was white with dust for weeks and barges were stranded in the upper Mississippi River. Yields of corn and soybeans dropped by 20 to 30 percent. For the first time in its history, the United States produced less grain than it consumed.

That same summer, on June 23, as the temperature in Washington, D.C., climbed to a record 101 degrees, NASA scientist James Hansen stood up at a congressional hearing and became the Paul Revere of global warming. A complex computer model he had created of the planet's climate system revealed that all the extra carbon in the atmosphere was heating up the earth. "The greenhouse effect has been detected," he testified, "and it is changing our climate *now*."

Alarmed by the growing evidence, that same year the World Meteorological Organization and the United Nations Environment Programme launched what would soon become the largest collection of scientists ever brought together to study a single issue. Made up of thousands of climate scientists and other scholars from around the world, the Intergovernmental Panel on Climate Change (IPCC) set out to employ rigorous, peer-reviewed scientific study to assess the fast-unfolding phenomenon of global warming.

Not that most people across the planet needed someone with a Ph.D. to tell them that something weird was happening to the weather. By the mid-1990s, the arctic had warmed so much that native Inuit people were at a loss for words—literally. Their one-thousand-year-old language and oral tradition had no word for the hornets, barn owls, warm-water fish, and other species suddenly appearing in their midst. No one had ever seen such creatures before. On the other end of the world, at McMurdo station in Anarctica, it actually rained in the summer of 1997, an event so rare it's the weather equivalent of snow in Saudi Arabia. In between these two poles, from the shrinking glaciers of Peru to the record heat waves in India to the dying coral reefs in the Caribbean, the evidence kept piling up throughout the 1990s.

Then came a final bombshell. In January 2001, after a dozen years of study, the prestigious Intergovernmental Panel on Climate Change served notice that the global climate system was hurtling toward major upheaval. In its now-famous "Third Assessment Report," the IPCC chronicled in three thousand pages an astonishing range of carefully observed and scientifically measured impacts from warming. The panel also ruled out all the culprits responsible for the earth's many climate shifts in the past, including volcanic activity and changes in radiation from the sun.

The only possible source of most of the observed warming, the panel concluded, was human combustion of oil, coal, and natural gas. Or, in science speak, "an increasing body of observations give a collective picture of a warming world" and "there is new and stronger evidence that most of the warming . . . is attributable to human activities."

Amazingly, the panel determined that all the profound climate disturbances observed by 2001 were the result of just one degree of warming since 1900, with most of that warming occurring in the last fifty years. More disturbing still, the panel predicted that if world fossil fuel use in the twenty-first century continued to rise as projected, the earth would experience an additional warming of between three and ten degrees Fahrenheit by the year 2100.

Now, just for a moment, let those numbers sink in: three to ten degrees of warming. To the average person that might not sound like much. Indeed, it might seem like a good thing. I'm from Georgia. I *hate* cold weather. My idea of a perfect place to live is where you know it's winter only when the mangoes fall off the trees. What could possibly be wrong with longer summers, longer growing seasons for farmers, less winter, less ice and snow?

Nothing, until you realize that even very small changes in the earth's heat budget will bring enormous changes to the global climate. Again, all the changes we're now seeing—record droughts, floods, glacier melt—are caused by just one degree of warming. What would happen with three to ten degrees more warming?

To give you an idea, consider this: Fifteen thousand years ago, toward the end of the last Ice Age, there was, over what is today New York City, a glacier so huge it measured a mile thick from its base to its top. It was a massive sheet of ice taller than four Empire State Buildings stacked on top of one another.

Today, of course, there *is* no glacier over New York City. We have a completely different global climate regime. But exactly how many degrees of warming occurred between the climate system then and now? Twenty-five degrees? Fifty degrees? No. The warming has been somewhere between a measly five and nine degrees Fahrenheit. That's all. Just five to nine degrees of warming is all that separates a global climate regime that creates a mile-thick glacier over New York City and one that gives us the Big Apple of today with the green grass of Yankee Stadium and horse-drawn carriages through Central Park.

Now scientists tell us we could get a full *ten* degrees of warming, occurring not over thousands of years as has happened in the past, but crammed into the eyeblink of one hundred years. That world, just a few decades away, arriving largely in our lifetime, would look nothing like the world we presently inhabit. *Nothing.*

Even a warming of "just" three degrees, on the low end of the IPCC's projected range, would bring real misery to human communities and ecosystems worldwide. Entire Pacific Island nations would disappear from sea-level rise. Malaria would continue to expand its range across the planet, including a probable significant return to the United States. And in Maryland, where I now live, credible projections warn that even moderate warming could reduce agricultural yields by up to 40 percent.

Can this *really* be true? Such scenarios are so extreme they demand careful skepticism equal to the alarm they arouse, right? Most of this, surely, is just the usual fear-mongering of environmentalists or the hyperbolic work of anti-American foreign scientists involved in the IPCC process.

Such were the sentiments, at least, of many top Bush administration officials soon after the president first took office in January 2001. Given that virtually everyone in charge at the new White House hailed from long and lucrative careers in the fossil fuel industry, including the president himself, it's not surprising that one of Bush's first acts in office was to try to cast doubt on the IPCC's findings. He asked America's premier scientific body, the National Academy of Sciences (NAS), to review the findings of the IPCC's two-thousand-plus scholars and get to the bottom, once and for all, of this "three to ten degrees of warming" business. The president was careful, moreover, to hand the issue to an NAS panel that included several leading "skeptics" intent on refuting the growing data of climate change.

But in June 2001, this same panel handed back to the president an answer he didn't want to hear: It's all true. Yes, the planet is warming, the panel said. Yes, human beings are the main drivers. Yes, the IPCC's prediction of three to ten degrees of warming by 2100 represents expert consensus based on "admirable" science.

The NAS corroborative review "should finally put to rest any unwarranted doubts about the reality of human-induced global warming," said Dr. Ralph Cicerone, the noted University of California–Irvine atmospheric chemist and current NAS president who headed the White House–commissioned effort.

But enough with the numbers and all the scientific talk. What does this really mean for you and me? What's the bottom line? It turns out you don't have to analyze the latest computer models from the National Oceanographic and Atmospheric Administration (NOAA) to get a clear picture of the super-hot world now forecast by leading climate scientists. All you have to do is go to the great state of Alaska right now. If the sky has already fallen on New Orleans, it is rapidly tumbling to earth upon the people and landscapes of Alaska at this very moment thanks to global warming.

THE POLAR REGIONS of the earth are heating up several times faster than the rest of the planet. That's because when ice and snow melt they take with them a white reflectivity that previously bounced sunlight back into space. The newer, darker surfaces of exposed water and soil absorb sunlight instead and hold its heat. This "feedback loop" greatly amplifies the local warming. So while the planet as a whole has already seen an average warming of one degree, Alaska has seen a rise of five full degrees just since the 1970s. The result has been an unfolding ecological and cultural disintegration not unlike coastal Louisiana prior to Katrina. And as with prestorm Louisiana, few Americans outside of Alaska are even vaguely aware of the emerging calamity there or its implications.

In Alaska today, there is a standing forest of spruce trees so vast it's bigger than the state of Connecticut—and yet every single spruce is dead. Beginning just south of Anchorage, you can drive and drive for hours and hours and see nothing but the ruined, rotting, dull-gray remains of these millions of century-old trees, stretching to every horizon, covering every hill, filling every valley floor. Across four million acres this haunting graveyard spreads, every tree destroyed by a single insect species: the Alaskan spruce beetle. With record mild winters the past twenty years, the beetles now thrive, reproducing at twice their normal rate, their numbers simply overwhelming the trees. The result is the largest forest die-off by insect infestation ever recorded in North America. And federal researchers give direct credit to the warming climate.

Alaskan glaciers, meanwhile, are in full-blown retreat. Columbia Glacier, one of the state's largest, is melting so fast it's retreating seven feet per day up the mountain slope. Since the 1950s, the state's glaciers have collectively shed about 9 percent of their mass per decade. That's about 2,000 cubic miles of melted ice and snow, creating enough liquid to turn the entire state of Texas into a New Orleans, covering it with tens of feet of water. Imagine the Astrodome and all of Houston inundated. Dallas under water. The northern plains, the western desert, the central hill country—all flooded like the Ninth Ward.

In the vast region north of Fairbanks, much of Alaska's land is simultaneously buckling, sinking, dissolving, cracking, crumbling, and splitting apart. The permafrost, which for thousands of years kept the soil frozen up to two thousand feet below the surface, is now thawing. It's no longer "permanent." As a result, roads all across the region now heave and dip and break apart from the suddenly mushy, spongy foundation below. Telephone poles tilt and "drunken forests" now abound in which trees lean at wild angles. And thousands of hydraulic jacks help prop up houses, previously stable for generations, that would otherwise see a dropped tennis ball roll and roll from room to room.

"It's kind of like living on a giant water bed," one Alaska resident told National Public Radio concerning the softening, undulating terrain that no one now alive has ever seen before.

Other impacts range from the anecdotal—Alaska's Iditarod dog race had to be moved hundreds of miles north in 2001 due to lack of snow—to the monumental—the eight-hundred-mile-long Trans-Alaska Pipeline is now at risk from global warming. *New York Times* reporter Timothy Egan wrote in 2002:

> *Engineers responsible for the pipeline, which carries about a million barrels of oil a day and generates 17 percent of the nation's oil production, have grown increasingly concerned that melting permafrost could make unstable the 400 or so miles of pipeline above ground. As a result, new supports have been put in, some moored more than 70 feet underground.*

"We're not going to let global warming sneak up on us," said Curtis Thomas, a spokesman for the Alyeska Pipeline Service Company, which runs the pipeline. "If we see leaning and sagging, we move on it."

The resulting cycle, of course, is as surreal as the melting landscape: We burn oil, which warms the planet, which melts the tundra, which causes the oil pipeline to sag, which requires major reinforcements, which allows us to burn more oil, which warms the planet more, which causes more tundra to melt, which . . .

The changes in Alaska are not restricted to land, of course. Beyond the state's northern shore, the frozen Arctic Ocean is melting away like ice on a Miami sidewalk. Just since 1979, satellite data show that roughly 250 million acres of perennial sea ice has vanished. That's a loss equal to subtracting from the lower forty-eight states an area five times the size of New England. In 2005 we saw the biggest decline yet: 70 million acres of sea ice gone in one year. That's another New England plus New York state in twelve months!

Alaskan polar bears, confused by this altered world, now struggle desperately to swim between vanishing ice floes. But increasingly their massive legs tire out and their mighty lungs, capable of ground-shaking roars, fill up with cold arctic water. They drown. For the first time any scientist or native hunter can recall, the polar bears of Alaska are drowning.

Researchers found four of the 700-pound bears floating dead in open ocean water 60 miles north of Barrow in September 2004. As many as 36 others may have died as the bears apparently tried to swim south to solid land after the arctic ice shelf receded an astonishing 160 miles to the north. Polar bears are as strong in the water as any land-based mammal, having adapted to swimming long distances. But they can't swim 160 *miles*.

Consequently, as a species, the bears are now headed for steep declines, if not extinction, in coming years, according to scientists. The latest projections are for arctic summers that are totally free of sea ice—*totally*—by 2100. After a hundred thousand years of living on the ice floes and hunting seals, the noble polar bear will see its habitat shrink and melt away and vanish completely in my son's lifetime.

AND THEN THERE ARE THE NATIVE PEOPLE of northern Alaska, the ancient Inupiat. Like the polar bears, they are also disappearing into the sea. Their coastal villages are now assaulted by ocean waves, their traditional food supply is in tatters, their drinking water is turning salty, and some elders say the end of time is near.

The breathtaking retreat of arctic sea ice has triggered unprecedented erosion all along Alaska's north shore. Storm waves and tides now pound the fragile coast, unhindered by the ice chunks that once played the equivalent role of barrier islands and wetlands, buffering the shoreline.

The tiny Inupiat village of Shishmaref, on the Chukchi Sea, exemplifies the problem facing dozens of indigenous communities. Three homes have fallen into the sea in recent years, seven have been relocated, and state engineers believe the whole community of six hundred houses could be swallowed by the advancing water within the next decade or so. Already, sea waves sweep away drying racks for fish and game while Shishmaref's drinking water has been contaminated by saltwater. Their tiny airstrip, their lifeline to the world, is just a few feet from a precipitous bluff that crumbles more every day.

For several centuries people have lived in Shishmaref. Now, a rise of five degrees Fahrenheit is making it uninhabitable.

No wonder the end seems near for so many native Alaskans. In 2001, the Inupiat people of Barrow, living in the northernmost human settlement on mainland North America, looked up to see the first thunderstorm ever recorded there. Ring-neck ducks, virtually nonexistent prior to the mid-1980s, are now common in northern parts of the state. And tuna were recently spotted in the Arctic Ocean for the first time ever.

But it's the changing pattern of whale-hunting that has had the biggest impact—symbolic and practical—on many Inupiat people. The whale is to the Inupiat what the buffalo was to the Plains Indians: a source of food, tools, folklore, and overall cultural identity. Even with the arrival of modern influences, the harpooning of whales has remained an essential way many Inupiat see themselves fitting into the world.

But the migrating bowhead whales now arrive up to forty-five days earlier in the year, totally altering the annual rhythms of Inupiat life. And the disappearing ice floes that once sustained so many polar bears also provided hunting grounds for Inupiat whalers tracking their prey out to sea. Now, even if a hunter finds and harpoons a whale, there's no solid ice platform onto which the massive catch can be hauled, butchered, and carried away by sled. And dragging an unbutchered sixty-ton bowhead to shore through so many slushy miles of soft ice and water is impossible. Fewer whales are being killed as a result, and the tradition is showing inevitable signs of decline.

"The people of the arctic are an endangered species," says Sheila Watt-Cloutier, president of the Inuit Circumpolar Conference, an alliance of 155,000 indigenous people that includes the Inupiat. "Everything we know, every part of our life, is coming apart."

The demise of whale hunting, the physical erosion of whole villages, the weekly news that even expert elders now head out across the unpredictable ice and never return, drowning tragically like the polar bears—these are all signs of a society dying right before our eyes. It is a "collapse" in the Jared Diamond sense. Not because of war. Not because of adverse changes in basic commerce. And not because the Inupiat have abused and overtaxed their local environment as happened on Easter Island and in coastal Louisiana.

No, the single overwhelming force exterminating this society is the last phenomenon on Diamond's list of killers: climate change. Global warming and nothing else is erasing the Inupiat.

◆   ◆   ◆

MY GRANDFATHER WAS BORN during World War I, got married and had kids during World War II, and died while America was embroiled in the Vietnam conflict. Late in his life, weary of all the twentieth-century bloodshed, he declared that there was no hope for the human race. Nothing, he said, would ever steer the world's people toward genuine cooperation and away from the ancient tendency toward war and mutual hatred. Nothing, that is, except one of two things: the arrival from outer space of aliens with hostile intentions, or the approach of a meteorite in need of being shot out of the sky. Only clear outside threats like these, he said, threats to the planet as a whole coming from outside our world, would ever unify us in a common, cooperative purpose.

But my grandfather was wrong. He died before the phenomenon of global warming was widely understood. We don't need aliens or meteorites attacking our planet to bring us together. We've attacked the planet ourselves, disrupting our one and only climate system. Now, whether we like it or not, we're all finally on the same team. The Wall Street broker and Calcutta street child and Australian barley farmer all share a clear and vital common interest.

Let's be clear about this: What's happening to the Inupiat in Alaska today is about to happen to all of us, everywhere. These native people are just the forerunners. Their region is warming faster than yours and mine. When temperatures eventually rise by three or five or ten degrees everywhere, all across the globe, as per the widespread predictions, then food production will unravel *everywhere*. Homes and infrastructure will come under assault *everywhere*. Civilization will be in upheaval *everywhere*. It's just a matter of time.

"My arctic homeland is now the health barometer for the planet," said indigenous leader Watt-Cloutier. "If you don't like what you see here, fix the problem before it spreads to everybody on the planet."

And, thankfully, there *is* a way to fix the climate problem. It just won't be easy. That's because no single tribe or city or state or country or continent is causing the problem. We're all causing it. So every tribe, city, state, country, and continent must be part of the solution. We're on the same team.

But being on the same team—by necessity, if not by choice—is not the same as playing well together. And so far we're not looking so good out there on the global field.

Our climate is changing because every day human beings around the world set fire to 15.9 million short tons of coal. We simultaneously combust 82.4 million barrels of oil and put a match to 7.4 billion cubic meters of natural gas—day after day after day. To repair and protect our life-giving climate, we have to stop doing this. We have to switch to energy sources that don't generate carbon dioxide.

And here there's good news. The Intergovernmental Panel on Climate Change, the same prestigious group of scientists that predicts up to ten degrees of warming by 2100, also makes clear that we still have enough time to slow down the warming process and perhaps ultimately stop it. What's needed is a roughly 70 percent reduction in current carbon dioxide emissions worldwide in the next few decades.

That's a huge drop, requiring revolutionary change in energy use. But there's more good news. According to the IPCC, we don't need blind luck or divine intervention to make the revolution happen. We don't need scientists to scramble to invent some previously elusive magic energy bullet. All the technology we need already exists and is presently in limited use, in one form or another, in modern wind farms and solar panels and hybrid cars and hydrogen fuel cells and biofuels.

What's needed is simply the global political will to bring these technologies to widespread use everywhere, among all people, as soon as possible.

The first major step in this direction happened in 1992, just a few years after James Hansen played Paul Revere before the U.S. Senate. Meeting in Rio de Janeiro at the so-called Earth Summit, leaders from 172 nations, including George H. W. Bush, negotiated the United Nations Framework Convention on Climate Change (UNFCC). Signatory nations pledged to adopt policies and take actions that would result in a "stabilization of greenhouse gas concentrations in the atmosphere at a level that would prevent dangerous anthropogenic"—manmade—"interference with the climate system."

Five years later, sorting out the nuts and bolts of a specific rescue plan under the UNFCC process, eighty-four nations, including the United States, signed the Kyoto Protocol. This document requires industrial countries to cut their aggregate CO_2 emissions by 5.2 percent below 1990 levels by 2012.

But faced with ferocious resistance from U.S. oil companies and automakers, President Bill Clinton never submitted the Kyoto Protocol to the U.S. Senate for ratification. He hoped instead to pick up the additional Senate votes he needed before leaving office. It never happened.

Then came George W. Bush. The former oil executive and ex-governor of oil-rich Texas arrived in Washington with an explicit agenda: Protect the fossil fuel industry at all cost. Without bothering to wait for his National Academy of Sciences panel to report back to him on climate change, the president shocked the world by immediately withdrawing the United States from the Kyoto process altogether. Henceforth, global warming would be referred to at the White House only as an "issue," not a "problem." It would be "studied," not acted on.

None of this is very surprising given that the new president and vice president and national security adviser and commerce secretary and White House chief of staff and scores of deputies and presidential advisers all came from oil or automobile backgrounds. Suddenly the oil industry no longer had to lobby the White House. The oil industry *was* the White House.

So there would be no reduction in fossil fuel use in America. Quite the opposite. The president endorsed an energy policy committing the nation to dramatically *more* fossil fuel use—more coal, more petroleum—and with tens of billions of dollars of taxpayer subsidies to the industry to make sure it happened.

All of this despite the avalanche of alarming data coming from the NAS and IPCC and other unimpeachable scientific sources worldwide. And despite the fact that even Pentagon planners had begun exploring military contingencies for the economically unstable and socially violent world of climate chaos just ahead. And despite the fact that at least one Bush cabinet member, Paul O'Neill, head of Treasury, openly compared the projected impacts of climate change to a "nuclear holocaust." And despite the fact that Britain's chief science adviser warned that global warming was already a bigger threat to mankind than terrorism.

Despite all of this, the United States, with less than 5 percent of the world's population but with 25 percent of its greenhouse gas emissions—the United States would leave it to Japan and the nations of Europe to begin desperately trying, on their own, to move forward with the Kyoto process.

We, meanwhile, would strike a decidedly Easter Island pose. Future anthropologists will surely recognize all the old patterns: the illusion of permanent prosperity, the obsession with short-term gain over long-term survival, the conflicting interests within society (oil welfare versus everyone else's welfare), and the leaders who were completely insulated from the problem. (As record heat waves hit Texas in the early 2000s, George Bush equipped his Crawford ranch house with a sophisticated and powerful geothermal air-conditioning system that kept everyone inside nice and cool despite the sweltering weather.)

The planet, of course, was wholly unconcerned with all these complexities of human psychology and greed. Per the physical laws of the universe, it just kept right on changing. Eighteen of the nineteen warmest years ever recorded have now occurred since 1980, and 2005 just set the record as the warmest year ever, according to NASA's Goddard Institute for Space Studies.

As U.N. Secretary-General Kofi Annan likes to say, "The climate is speaking back."

It's speaking back in several ways, actually, creating a line of communication that gives loudest voice, not surprisingly, to the world's oceans. The earth is a water planet. Hydrogen oxide, as liquid or as ice, covers roughly 75 percent of the globe's surface while, in the form of rain or snow, it falls intermittently over virtually every square inch. So the principal effect of altering our global climate is to rearrange, on a massive scale, the quality and movement of every cubic inch of water on earth. This means precipitation patterns change. More droughts occur here. More floods there. And icebergs melt. Glaciers retreat. Seawater becomes less salty. Ocean currents speed up or slow down. And the air becomes warmer, holding more moisture. The world's water is completely rearranged.

But the biggest "water change" of all is as straightforward as it is devastating: Warmer water takes up more space. It swells in volume. It's that simple. And therein lies the issue that connects more human beings to the potential harm of global warming than perhaps anything else. The world's oceans, which already cover most of the planet, are getting warmer, so they're getting bigger. They're jumping past their old boundaries and they're on the move. And where they're moving is directly toward us, the human species.

Native Alaskans and polar bears aren't the only ones who'll be drowning soon. George Bush and the oil lobby are triggering a mass uprising—of water.

5

Sea-Level Rise: Exporting New Orleans to the World

OF THE MANY LESSONS emerging from the demise of New Orleans, perhaps the most important is one scientists call the "relative sea-level rise." The ocean, it turns out, can move closer to a coastal area in three different ways: (1) the sea itself rises to a higher level; (2) the land sinks toward the sea; or (3) the two happen at the same time.

In much of coastal Louisiana over the past century, the land sank about two full feet. At the same time, the Gulf of Mexico, as part of a worldwide trend, rose about a foot. The combination of these two factors created a "relative sea-level rise" of three full feet. The Gulf of Mexico climbed that far up a ladder toward the city of New Orleans.

The result, as we've seen, was catastrophic. A million acres of buffering wetlands literally drowned and disappeared. A vast network of tall and broad barrier islands were reduced to skeletal wisps of land or were extinguished entirely. And the final result, in 2005, was Katrina, riding in on a runway of open water, crashing headlong into the heart of New Orleans.

It'd be nice to think that this was an isolated case, a phenomenon peculiar to poor old New Orleans, the town below sea level. The rest of us, safe in our own coastal communities, could simply shake our heads and write a donation check and go about our business. The only problem with this response is it ignores the stubborn reality of that first category of ocean attack: when the seawater itself rises up.

Among the many disturbing impacts projected by the Intergovernmental Panel on Climate Change is this one: Global warming will cause the world's oceans to rise somewhere between one and three feet above current levels by 2100. These are astonishing numbers. Let them sink in for a moment. Every coastline in the world, in other words, may soon find its relationship to the sea altered a full three feet in favor of the ocean, *just like New Orleans.*

Again, the National Academy of Sciences, in its 2001 report to the president, endorsed the IPCC's findings, including those on sea-level rise. And the Bush administration's own scientists and policy makers have since 2001 embraced these same numbers—one to three feet by 2100—in official reports and letters.

That glad-it's-not-me feeling you've had toward post-Katrina New Orleans should be fading by now. If you live anywhere from lower Manhattan to inner-city Baltimore to the southern suburbs of Houston you now know what it was like to be a resident of New Orleans prior to August 29, 2005. Major, major change is coming your way and it won't be pretty.

Louisiana's towering three feet of relative sea-level rise was mostly caused, as we've seen, by the land subsidence brought on by the strangling flood levees along the Mississippi River. But for the rest of the world it's our cars and gas furnaces and dirty electricity and petroleum-intensive agriculture that's causing the problem, driving up global warming. That warming in turn is causing the world's oceans to rise in two major ways.

First, of course, land-based glaciers are melting at breakneck speed, not just in Alaska, but worldwide, and the resulting meltwater is flowing to the sea. Glacier National Park in Montana had 150 glaciers spread across its rugged peaks in 1910. Today, with steadily rising temperatures, fewer than thirty glaciers remain and the park will be entirely free of perennial ice by 2030, prompting speculation that the park will have to change its name.

Mount Kilimanjaro in Africa, meanwhile, was once a great white castle of snow and ice rising above the plains of Tanzania and Kenya. In his famous short story "The Snows of Kilimanjaro" published in 1938, Ernest Hemingway wrote, "As wide as all the world, great, high, and unbelievably white in the sun, was the square top of Mount Kilimanjaro." Today, almost every bit of that cold, white landscape is gone, melted away, vanished. Kilimanjaro will be totally ice free within a decade, experts say, thanks to a warming trend that intensifies every year.

With few exceptions, such scenes are being repeated across every rocky slope on every mountain range across every continent on earth. And with time, virtually all of that meltwater reaches the ocean. Scientists, in fact, now estimate that an astonishing 118 trillion gallons of water—previously locked up as continental ice—streams into the oceans worldwide each year, swelling its volume. That's equal to all of Lake Erie flowing into the ocean every twelve months.

Of course, land-based glaciers are not the only ice that's melting on the planet. Polar sea ice is vanishing, too, as we've seen. Recent spectacular examples include the Larsen Ice Shelf B in Antarctica. In the spring of 2002, this frozen monolith the size of Rhode Island completely collapsed and disintegrated over a thirty-one-day period due to warming water from below and warming air from above. And in September 2003, the Arctic Ocean's largest ice sheet, the Ward Hunt Ice Shelf, broke in two and collapsed completely. It was 3,000 years old, 80 feet thick, and 150 square miles in area.

But whatever the impacts these events may have on polar bears and native people, the melting sea ice is *not* contributing to global sea-level rise, as many people believe. That's because water has the same displacement value whether it's frozen or liquid. That full glass of iced tea you forget about on a summer porch, after all, hasn't overflowed by the time you return to find all the ice melted.

So it's land-based ice that's driving sea-level rise. But even this source of water, as great as it is, ranks second to another phenomenon. Like the air and the land, the world's oceans are themselves warming up. And when water gets warmer, the natural vibration of the water molecules intensifies. As the heat energy grows, these vibrations become stronger and stronger and each molecule is pushed a little farther apart from the other and the volume of the water increases. Scientists call this thermal expansion. And when untold zillions of molecules of saltwater do this, the result, quite simply, is a bigger ocean.

Best estimates are that two-thirds of the projected sea-level rise for the twenty-first century will come from thermal expansion, the rest from meltwater. The mean surface temperature of the world's oceans has been rising steadily for decades and shows no sign of slowing down, of course. Tellingly, the surface water in much of the Gulf of Mexico was two to three degrees above average just prior to Katrina's arrival.

What concerns scientists even more, however, is the fact that this warming is now radiating downward, reaching as far as two miles below the surface to the deep oceans. This greatly enlarges the amount of water subject to thermal expansion and explains why the IPCC, representing the largest scientific collaboration in human history, now forecasts up to three feet of ocean rise by 2100.

These numbers bring to mind the ancient Chinese saying, meant as a curse: May you live in interesting times. For the three billion global inhabitants who live near a coastline, this will be a very interesting century indeed.

RELYING ON SPECULATIVE, highly uncertain evidence about weapons of mass destruction that never existed, the U.S. government sent over 100,000 troops to Iraq in 2003 and now spends $9 billion per month to keep them there, all in the otherwise worthy cause of protecting our homeland from foreign assault.

But when a mountain of published, peer-reviewed data of the highest scientific integrity shows that America is about to permanently lose up to fourteen thousand square miles of sovereign territory to an invading force, we do nothing to defend ourselves. The stark facts of global warming are that our coastlines are about to crumble, dissolve, and retreat just about everywhere. Not only has our own National Academy of Sciences validated the projection of up to three feet of sea-level rise by 2100, but so have the academies of science of seventeen other nations, including Canada, Brazil, and most of Europe.

Fourteen thousand square miles of U.S. land will turn to water as a result. That's an area slightly bigger than Delaware and Connecticut combined. Or, instead of the ten square blocks of Manhattan destroyed on September 11, we'd lose an area big enough to hold forty-four entire New York Citys. Lower Manhattan itself, including the very site of the former World Trade Center, would be flooded during even modest storms.

Other countries are in for an even rockier ride. Bangladesh will lose 15 percent of its entire landmass to the Indian Ocean by 2100 from three feet of ocean rise. Tragically, an even higher percentage of that nation's food comes from this death-row coastal area. Holland will lose 6 percent of its land area to the North Sea. Much of Egypt's densely populated Mediterranean coast will vanish. And small-island nations like the Maldives will disappear almost completely. Seventy-five percent of the Maldives land area, spread across 1,200 islands, is 3.5 feet above sea level or less. During the 2004 tsunami, the entire country vanished for several minutes under the Indian Ocean, a foreshadowing of things to come. And already New Zealand has agreed to accept "climate refugees" from the low-lying island nation of Tuvalu, in the South Pacific, when the inevitable end comes. Concrete plans are being made *right now*, in other words, to evacuate whole countries.

To comprehend the full scope of the problem, consider that thirteen of the world's fifteen largest cities are on coastal plains, including

Calcutta, Dhaka, London, Lagos, New York, Shanghai, and Tokyo. Obviously, all you have to do is look at coastal Louisiana over the past century to see what's coming: serious land erosion, saltwater intrusion into drinking water, disruptions to local fisheries, increased exposure to storm surges, and, unless massive seawalls and other "armored" infrastructure are constructed, displacement of hundreds of millions of people per the New Orleans example.

The impact of sea-level rise on the world's plant and animal species is, likewise, almost too great to comprehend. The demise of the world's coral reefs alone will be a significant blow to regional economies and to the global ecosystem. These brilliant, colorful "rain forests of the sea" are more than just playgrounds for the multitrillion-dollar tourism industry. Like wetlands, they protect tropical coastlines from surge tides. And a hefty 10 percent of all commercially important marine fish worldwide are dependent on coral reefs at some point in their life cycle. This is 20 to 25 percent of marine fish consumed by poor nations. Without the reefs, these catches will decline sharply or disappear altogether.

Corals grow very slowly by nature and simply can't keep up with fast-rising seas. To fall several feet below the ocean surface reduces the amount of penetrating sunlight that is needed to sustain the colonies of microscopic algae that are the building blocks of a reef. Warmer water also "bleaches" the coral, killing the algae. Seventy percent of all reefs in the Indian Ocean were rendered essentially comatose during the record-warm year of 1998 alone. A startling debate has now broken out among many marine scientists as to whether it's already too late to save the world's ancient coral reefs from the ongoing and projected impacts of global warming. They could all be gone in our lifetime if present trends continue.

Along with this unfolding ecological bankruptcy, Britain's largest insurance company, CGNU, in 2002 predicted that unchecked global warming could bankrupt the entire global economy by 2065. A key threat highlighted by the insurer was sea-level rise that would directly destroy valuable land, buildings, and agricultural assets while indirectly exposing everything farther inland to the more intense storms expected in a warmer world. A huge share of human infrastructure would simply become "unprotectable" and therefore uninsurable and therefore unavailable for meaningful economic activity.

LIKE MOST PEOPLE, my brain grows numb thinking about global warming in terms of ice caps melting and Pacific island nations disappearing and vast coral reefs dying. The evidence is shocking but also very far away from my home. Maybe if polar bears roamed my state of Maryland, their graphic death by drowning would help focus my mind on global warming in a daily way. But they don't live near me.

My friend Mike Tabor, a farmer, however, *does* live near me, and he's drowning, too. His sixty-acre farm just outside of Hancock, Maryland, has seen two hundred-year floods and four fifty-year floods just in the past ten years. In fact, watching Mike's farm gradually die at the hands of freakish, unprecedented weather is pretty much all I need to know about global warming. Climate change is, by definition, a local problem for everybody, manifested everywhere. It's not a distant abstraction. The signs are all around us, local signs, and if we look for them we'll not only find them but find they are intensifying with great relevance to our present and future lives.

If this were some futuristic, interactive book I would have you press a button right here to read twenty pages about all the observed and projected impacts of climate change right where you live. Rocky Mountain states? Read how the ski industry is struggling to attract customers and make artificial snow as warming temperatures threaten to wipe out the whole industry. Aspen will likely have the climate of Santa Fe, New Mexico, by 2100 if present trends continue. Great Lakes region? Read how these "inland seas" are actually *dropping* in water level—with potentially huge impacts on industrial harbors and overall shipping—because the lakes spend less time each year frozen and more time as evaporating water. New England? The treasured maple syrup industry is highly endangered as sugar maples migrate toward Canada to outrun the warming weather. Midwest? Your tornado season is getting more intense. Southwest? Wildfire damage is growing thanks to hotter, drier conditions. (Visit www.climateemergency.org for more on regional impacts.)

But this isn't an interactive book and I don't have the space for a dozen detailed regional analyses of climate impacts. So I'll just "randomly" select one U.S. region to use as a case study. I'll select my own region. I live in Takoma Park, Maryland, just outside of Washington, D.C., on the southern edge of the great Chesapeake Bay watershed. I intend no bias or manipulation by choosing my own area. The point is simply to illustrate that while the details may differ where you live, you and I share the exact same story.

WHEN MIKE TABOR first began farming in the early 1970s along a narrow mountain valley ninety minutes northwest of Washington, D.C., the weather patterns were highly predictable. First frost in mid-October. Last frost in early May. Plenty of rain for April/May planting. Hot and dry in August. Rains returning in September or October.

"But today I have no idea at all what to expect from month to month or year to year," says Tabor, sixty-three. "I'll get a freak killing frost in late May and then a first autumn frost in November. And the rain! Totally unpredictable. One year I'm under water and can't work the fields until late May. The next year a record drought. Last year my creek went completely dry for the first time ever. Never happened before."

There's no way to prove absolutely that the weird weather in Mike Tabor's area is directly related to global warming, but it sure fits the federal government's own projections. Because of the inevitable precipitation changes and the increased extreme weather events brought on by rising temperatures, the U.S. Environmental Protection Agency (EPA) predicts that acreage devoted to top agricultural crops in Maryland and Pennsylvania could drop between 39 and 43 percent in the coming decades.

Right now I get many of my summer vegetables from Tabor, hauled down to D.C.-area farmers markets. And I buy my Christmas tree from his farm each year.

"But I'm thinking of quitting farming for good," Tabor says. "Not because I don't love farming. I do. The weather's just making it harder and harder. The unpredictability is a major factor in my thoughts of early retirement. I just never thought it would end this way."

But you don't have to be a farmer to notice the odd weather in the Chesapeake region. For the past few years we've been getting thunderstorms in January and February in Maryland, something *none* of the many native residents I've queried can remember as children.

Equally weird, we got four major power outages over a ninety-day span in 2003, lasting anywhere from a couple of days to a couple of weeks. These were caused by a freak windstorm, two monstrous rains, and a tree-killing hurricane—Isabel—that spawned a record-high surge tide in the Chesapeake Bay to boot. For much of that summer/fall of 2003, we had the electricity reliability of a Third World country. The power companies said it was unprecedented.

It's not trivial to add that the fabled cherry trees in Washington, D.C., now bloom seven days earlier than in the 1970s, and that our own maple syrup industry in the Appalachian Mountains of Maryland and Virginia is largely dying out due in part to warmer, less predictable weather.

Of course, none of this is on the scale of Alaska . . . yet. There's no four-million-acre wasteland of dead trees or whole towns falling into the sea. But the impacts are multiple and noticeable and will intensify soon, according to Dr. Cindy Parker of the Johns Hopkins Bloomberg School of Public Health in Baltimore. Specializing in climate change and health impacts at the school's Center for Public Health Preparedness, Parker warns of many coming threats.

"Insects that carry diseases, like mosquitoes, could expand their range and bite more frequently under warmer conditions," says Parker. "Here in Maryland and across the U.S., we could see outbreaks of diseases we haven't had before, such as what happened with the West Nile virus in 2001. Globally, sharp increases in malaria and dengue fever are projected and could be devastating to populations that don't have strong public health systems to contain them."

But perhaps the biggest threat to U.S. health are the mega–heat waves forecast under future global warming. Already, heat waves kill more Americans each year than any other natural disaster, and the United Nations predicts a doubling of urban heat-related deaths in the United States by 2020. One 1995 heat wave alone killed over 750 people in Chicago. And no fewer than 35,000 Europeans died in the record heat wave of 2003, an event scientists now believe is directly linked to global warming.

But of all the climate challenges to agriculture in my region—and to health and commerce and the environment—none packs quite the same cumulative punch as the simple fact that the Atlantic Ocean, just down the road from where I live, is getting bigger. It's pushing water closer to my house and closer to my son's school and closer to the Maryland governor's mansion and closer to the bright white dome of the U.S. Capitol.

In downtown Washington, D.C., at the corner of 6th and E Streets, NW, stands a brand-new museum where you can push a button and see the future in vivid, awe-inspiring detail. The five-thousand-square-foot Marian Koshland Science Museum, operated by the National Academy of Sciences, has among its permanent exhibits an entire section on global warming. In the middle of the exhibit area, under a large sign that says "Climate Change Hits Home," hangs a giant satellite map of the Chesapeake Bay watershed. Below the map are two buttons. Push one and you see what happens to this region from roughly one and a half feet of sea-level rise. Push the other and you see what comes from about three feet of higher water.

I remember standing there one ironically warm winter day, about to learn the fate of my family and friends by pushing a button, and feeling decidedly uneasy. I remember hearing through the museum windows the busy sounds of Washington, D.C., on a weekday morning: cars honking and pedestrians talking and dogs on leashes barking at distant sirens. The city seemed too caught up in the calm of normal routine given the evidence of stark danger I knew this museum contained, evidence open to anyone who ducked in.

Off to one side, spread across an entire wall of the museum, was a giant graph showing the earth's very stable level of atmospheric carbon dioxide for the past ten thousand years. The graph then described the great carbon dioxide spike of the twentieth century and the almost vertical carbon takeoff projected for the twenty-first century. It was the famous "hockey stick," luridly splashed across a whole wall.

To illustrate what all that CO_2 means for the earth's temperature, there was a clever demonstration nearby featuring two halves of a large earthlike globe that visitors can touch. One half of the globe was covered with a thin layer of insulating plastic and was fed light from a bright lamp hanging just above it. The other half had a thicker layer of insulating plastic resting below a second lamp. Touch the thin-layer earth and it was warm. Touch the thick-layer earth and it was hot. The heat radiated into your fingertips.

But I was there to learn about sea-level rise in my backyard. For a moment I stood before the great satellite image of the Chesapeake Bay region, taking in the map's detailed features ranging from the barrier islands of the Atlantic coast to the headwaters of the Susquehanna River to the great cities of Baltimore, Annapolis, and Washington situated right smack in the middle of it all. It's a region featuring lots of water and marshes and farmland and sprawling suburbs and lots and lots of people. It seemed so orderly and permanent on the map. What a shame the atmosphere was changing.

I pushed the first button, the one simulating roughly 1.5 feet of sea-level rise, and watched as water poured in like a great tide across my neck of the world. Small green and blue lights flashed on, illuminating the water's path all across the map. All that meltwater from Montana glaciers, combined with the warming, swelling ocean waters, had to go somewhere—and here it was.

As the simulation proceeded, I suddenly had an overwhelming sense of déjà vu. I'd seen all this before, somewhere else. These images were far too familiar. And so was the wave of horror I experienced as the demonstration continued.

There, on the screen, a huge portion of my state of Maryland simply vanished in an instant. The entire lower third of Dorchester County on the Bay's Eastern Shore turned to open water. Historic Smith Island to the south, home to a vibrant community of oystermen and crabbers, disappeared almost completely as did Virginia's Tangier Island. Over on the Atlantic coast, Assateague and Chincoteague Islands were decimated by water while way up in Baltimore, in the upper Bay, suburbs were flooded. And everywhere in between: tidal rivers flowing into the Bay swelled into much larger and wider versions of their former selves.

The cumulative impact of all this was that the main stem of the Chesapeake Bay, not far from where the tidal Potomac River makes its way up to Washington and to within a few miles of where I live—this portion of the Chesapeake Bay nearly *doubled* in width. Tens of thousands of acres of dry land and marshes turned to open water, giving the overall impression that the Atlantic Ocean was moving into the neighborhood in a very serious way.

And this was from barely eighteen inches of sea-level rise. It's the "mid-range" projection of the IPCC.

I pushed the next button, simulating a full three-foot rise, and the devastation worsened. The Atlantic barrier islands were pretty much finished off. Flooding poured into Alexandria, Virginia, and the suburbs south and east of D.C., while the northern neighborhoods of Baltimore got still more water. To the east, the Nanticoke River swelled and widened so much it essentially cut the eastern part of Maryland in half. And the Chesapeake's crown jewel, the Blackwater National Wildlife Refuge, home to millions of resident and migratory waterfowl and often called the Everglades of the North, was completely wiped out. All 28,000 acres turned to open water.

This is what global warming means where I live. You don't have to go to a museum to see its early stages, either. The water level in the Chesapeake Bay near Baltimore rose about seven inches over the last century thanks to global warming. This, combined with a moderate amount of local land subsidence, forced the abandonment of thirteen inhabited bay islands in the twentieth century. Across the region, miles and miles of former farmland now lie idle from saltwater intrusion, and oystermen on Smith Island, sounding just like Louisiana shrimpers prior to Katrina, warn of being wiped out by a future hurricane thanks to all the disappearing marsh.

And this is just one U.S. coastal area. Just one example. A sea-level rise of one to three feet will, to a greater or lesser extent, impact every inch of American shoreline from the Texas coast to the Florida Keys to the Outer Banks of North Carolina to Cape Cod. The low-lying areas of San Diego and San Francisco and much of Puget Sound on the West Coast are at great risk, too.

A rise of just two feet in sea level will, according to the EPA, eliminate up to 43 percent of all U.S. coastal wetlands. And a study by the Federal Emergency Management Agency estimates that no fewer than one in four U.S. buildings within five hundred feet of a coastline will be destroyed by erosion by mid-century.

All manner of infrastructure will be flooded as well, creating a mess of potentially staggering dimensions. Landfills and hazardous-waste sites, neither of which can be picked up and moved inland, will be inundated, their toxins spread throughout the water. New York City, right now, has no fewer than fourteen wastewater-treatment facilities located exactly at sea level, just waiting for disaster.

At the same time that ocean water attacks the land, saltwater will seep into the groundwater below, penetrating aquifers like those in low-lying Maryland that are critical to farmland irrigation. Spraying crops with even slightly salty water is like spraying them with poison.

Philadelphia and much of California's Central Valley, meanwhile, draw their drinking water from sections of rivers that are only slightly upstream from the cutoff point where the rivers turn salty during droughts. Sea-level rise will push saltwater farther upstream from the mouth of rivers, meaning, of course, that drought conditions in the future could send brackish water flowing freely from downtown Philadelphia taps.

AND THEN THERE ARE THE STORMS. In September 1985, Hurricane Gloria steamed up the Atlantic coast and made landfall just above the mouth of New York Harbor, passing north of Manhattan. As a Category 2 storm, its surge tide could have been very serious indeed. But this storm was a relative dud, causing only minor flooding and spotty structural damage in beach communities across eastern and central Long Island. The reason? The New York area got lucky. The storm struck at low tide. It came when the ocean had conveniently lowered itself a full five feet in relation to the land, down from the high-tide mark of just six hours earlier. This created perfect conditions for a "soft landing." Had Gloria arrived at the peak of high tide, according to hurricane experts, this same storm would have been a pure monster, generating more flooding and more damage than even the Great Hurricane of 1938, the so-called Long Island Express. That storm brought up to six feet of water across Long Island, and the tidal wavelike storm surge killed more than fifty people and caused $15 billion in damage. Only the luck of low tide in 1985 saved America's largest urban region from its most serious hurricane threat in generations.

For every emergency preparedness official whose hair has turned white praying for similar good luck, praying for a storm to strike a coastal community at maximum ebb tide, global warming represents a state of permanent misfortune. By inducing up to three feet of sea-level rise worldwide, we are in essence creating high-tide conditions all the time, everywhere on the planet, guaranteeing that even Category 2 storms like Gloria will be surge-tide heavyweights.

Today, in Washington, D.C., it would take an extraordinary storm to seriously flood the national mall, located in the heart of downtown, just ten subway stops from my home. It would take what storm experts call a "hundred-year" surge tide. That storm would have to be so huge it could push a wall of water past the Atlantic barrier islands and past the Chesapeake Bay islands of Smith and Tangier and up and over thousands of acres of wetlands and up the Potomac River and finally over the top of the bulkheading along the Washington Shipping Channel and Tidal Basin into the center of the District of Columbia.

But what if you pushed that sea-level rise button at the NAS museum and suddenly, even before a hurricane hits, you've got a standing high tide so profound that there *are* no Atlantic barrier islands or Bay islands. The lower Bay itself has nearly doubled in width and the wetlands are all gone and every inch of remaining shoreline exists as if at the edge of a chronically full bathtub, where the Potomac River itself beats an always-swollen path right up to the gates of Washington, D.C. Under these conditions, hundred-year storm surges no longer happen every hundred years. They happen every fifty years, according to a report by the Pew Center for Global Warming Studies. With three feet of sea-level rise, a "fifty-year" storm surge would send saltwater pouring into the front door of the Air and Space Museum. In other words, you effectively double the frequency of hundred-year flooding in the nation's capital.

Other areas, due to unique local conditions, will have it even worse. Some U.S. coastal areas, according to the EPA, will experience hundred-year flooding every fifteen years if there's three feet of ocean rise. And none of these estimates takes into account the fact that tropical storms and hurricanes are themselves becoming more powerful with global warming, thus bringing to these already bad future conditions even higher surge tides thanks to more powerful wind speeds fostered by a warmer world (more on this in chapter six). The stronger winds work to push bigger and bigger tsunami-like walls of water into the land. Adding this factor means that monster hundred-year surges could very well strike Washington every twenty-five years or even every decade. Suddenly, the District of Columbia and nearby Baltimore and much of the rest of the region where I live starts to look a whole lot like New Orleans.

"You've got a crisis situation, certainly, for everyone who lives near the Chesapeake Bay," says Dr. Court Stevenson, a marsh ecologist with the University of Maryland. "The buffering landforms that are the enemy of hurricanes will be gone. And the high water that is the best friend of big storms will be here in abundance."

The same could be said for Miami, a city whose average elevation is barely three feet above sea level now; a city facing the potential loss of buffering coral reefs and keys to the south and the inundation and disappearance of much of the Everglades to the west.

And there's New York, a "water city" if ever there was one. It's composed of "just four little islands and a bit of mainland laced together by eighty bridges and tunnels—the entries to many of which are barely above sea level, if at all," writes *Sierra* magazine. Hurricane Gloria, like a blessing from heaven, landed near this mosaic of concrete, glass, and water at low tide, an option that essentially won't exist a few years hence.

"OH MY GOD, this can't be happening."

That's the thought that's bound to strike most people who push the dreaded Chesapeake "flood" button at the NAS museum. This thought is then likely followed by, "We've got to *do* something. We've got to fix this problem."

Of course there's little chance the nation's foremost scientific body would create an ambitious museum exhibit in the middle of the nation's capital if the problem wasn't really happening. It *is* happening.

And as for "fixing" it, all the options open to us are, unfortunately, fraught with pain and risk.

All except one, that is. The best way to slow and ultimately stabilize the rising oceans is to slow the atmospheric warming that is, itself, driving the process. And to do this requires an immediate and comprehensive conversion to clean, efficient energy as begun by the rest of the world under the Kyoto Protocol. In the United States, such a conversion would have the tremendous simultaneous benefit of enhancing national security, improving public health, and strengthening our economy, as we'll see in later chapters.

But President Bush has rejected Kyoto and all other viable strategies for switching to clean energy while offering no meaningful plan of his own even as he acknowledges that the warming trend is well under way. After all, the president formally asked the NAS to present him with much of the very same data now on full display at the Koshland Museum at 6th and E Streets. Without U.S. participation, the Kyoto process is unlikely to succeed fast enough to make a difference, leaving everyone in America with only the "pain and risk" options for dealing with sea-level rise.

While we wait for national action to address the warming, coastal land-use planners are stuck with two basic strategies—and only two— for confronting the rapidly rising ocean water. These strategies are "accommodation" and "planned retreat." In other words, you either get out of the way, or somehow you learn to live with the higher water.

Planned retreat is the least favored option, of course, since no one wants to be told to move back from the pretty shoreline. But many jurisdictions already require new coastal buildings be set back per the best scientific projections of long-term sea-level rise and erosion. Meanwhile, a great deal of retreat has already begun, ranging from the abandoned island communities in the Chesapeake Bay to the historic Cape Hatteras Lighthouse in North Carolina that was moved inland 2,900 feet in 1999.

But the preferred response to sea-level rise is accommodation, a euphemism for the more warlike language many planners use. They speak of "fortressing" a shoreline against "storm attack" and "water intrusion." This is done by "armoring" shores with bulkheading (retaining walls backed by landfill) and building "fortified levees."

In my state of Maryland, no fewer than three hundred miles of shoreline have been armored in recent decades. That's nearly 10 percent of the state's total coastline, with hundreds more miles likely to be hardened in the near future as sea-level rise accelerates.

But there are two obvious problems with this approach. First, it's very expensive. The cost of bulkheading a tidal shoreline in Maryland ranges from $600 to $1,000 per linear foot. The ultimate cost to this one state could be nearly a billion dollars, assuming the minimum cost of $600 per foot. Second, bulkheads wreck vital wetlands. The natural response of marshes is to migrate landward as the water rises. But increasingly, these grasses will meet walls of wooden pilings and concrete slabs that protect roads and homes and soccer fields and grocery stores. That's when the game is over. The wetlands vanish on a massive scale.

For some cities, full-blown levees may be an option for holding back the sea. But this, too, is very expensive, as New Orleans has shown. And we all know now that levees can break. While there's no doubt levees will play a protective role in certain parts of the country as the oceans swell, armoring every inch of U.S. coastline with massive levees or muscular bulkheads is neither geologically possible nor financially affordable. It just can't be done.

The instinctive American reaction at this point is to simply think bigger. The Chesapeake Bay is in danger? Why don't we just build a great sea wall across its mouth, thus protecting everything behind it in one efficient fell swoop? We'll re-create the Dutch "miracle" up and down all our coasts, saving the land through our engineering brilliance and sheer national will, leaving a legacy of adaptation to be studied and marveled at by later generations.

The truth, however, is that the Holland experience can't be replicated everywhere, nor would we want to. The mouth of the Chesapeake is too wide and there are no "hard islands" along the way that could anchor a seawall as in Holland. Such a wall would greatly disrupt the delicate estuarine life of the Bay anyway. Similar obstacles preclude Dutch-like seawalls to protect coastal Louisiana or the fragile shores of Miami.

Modified walls of modest distances might be viable in certain U.S. areas, true. But the more likely response would be so-called "flood-gates" for certain coastal cities in my region and across the country. Such gates would be narrower in width and laid across a tidal river or harbor entrance downstream from a metropolis. One might be built, for example, across the upper Potomac River near Mount Vernon, thus protecting D.C., and another across the Patapsco River along the upper Bay, thus protecting Baltimore. Unlike seawalls and levees, proponents envision these structures serving as doorlike barriers that can be opened and closed as needed in the face of storm surges. They would not interfere with shipping nor would there be the risk of flooding from heavy rainwater collecting behind the barriers.

Precisely this sort of structure already exists across the Thames River east of London. It was constructed after a North Sea storm in 1953 took three hundred British lives. And since construction, the barrier has been closed more than eighty times, serving effectively against surge threats.

Beyond the Chesapeake region, several U.S. cities might benefit from such structures, including Galveston, Tampa Bay, and Charleston. But New York City, with its 580 miles of tidal shoreline, stands to gain the most from such barriers. By erecting enormous gates at three strategic locations—the mouth of the Arthur Kill, The Narrows at the mouth of New York Harbor, and across an upper stretch of the East River—most of Manhattan and at least half the city as a whole could, in all likelihood, be kept safe and dry from future surge tides enhanced by sea-level rise.

The question, then, isn't whether many of these "accommodation" schemes will work. They probably will. The question is *how long* can they be made to work in the face of enormous uncertainty and risk. On what scale should we build protective structures given that we don't know precisely how high sea level will go or how much more intense storms will become? How many years of service should we plan for—fifty years? a hundred years?—and based on what assumptions? And who gets to decide, especially given that the bigger the storm you design for, the more expensive the project?

And then there's the ultimate question, posed recently by Klaus Jacob, an expert on disaster risk management at Columbia University. "Would building [flood] barriers simply postpone the inevitable: As sea-level rise accelerates, would we be setting a trap in which we feel safe behind barriers and continue to invest in areas that ultimately will be flooded?"

Are we really ready to become, all of us, New Orleanians, casting our fate behind ever-higher and suspiciously unsustainable walls? Once we commit to major warfare—fortified levees, armored bulkheads, massive floodgates—there's no turning back. It's an all-or-nothing proposition, as New Orleans has demonstrated in graphic detail.

This is the path of pain and risk. Unless we treat the disease of atmospheric warming, everything else is a costly crap shoot. Unless we kick our habit of fossil fuels, thus reducing the ocean climb itself, we'll be forever condemned to deal with the intensifying symptoms of sickness until, in all likelihood, countless communities and their brave, determined inhabitants will suffer the horrifying fate of New Orleans.

BUT DOESN'T UNCERTAINTY CUT BOTH WAYS? Maybe we'll get lucky. Maybe nature will surprise us or we'll make just enough energy changes in just enough time to keep the sea-level rise to the lower end of the projected one- to three-foot range. Twelve inches, after all, isn't that much more than the global average over the twentieth century, and we survived that, didn't we?

Yes, perhaps. But a great many scientists believe this is the least likely scenario of all. Even if, starting tomorrow, we never burned another BTU of oil, coal, or natural gas ever again, we'd still likely get a foot of sea-level rise. That's because carbon dioxide lingers in the atmosphere for up to a century after it's been released, doing its greenhouse work. Our best hope is to make a *rapid* switch to clean energy in the next ten to twenty years and hope to keep the atmospheric warming well below two degrees and the sea-level rise to maybe 1.5 feet.

In this book, however, I've focused mostly on the threat of three full feet of sea-level rise because it seems wisest to discuss the more serious end of the spectrum and to plan accordingly while letting the higher number perhaps motivate us to keep it from happening.

Yet amid all the unanswered questions about the precise extent of future impacts, one fact is fast becoming disturbingly clear. If we do nothing at all to reduce greenhouse gas emissions as is now official U.S. policy, we could wind up with *more* than three feet of sea-level rise by 2100. Much more.

This is the warning now coming from none other than James Hansen, director of the NASA Goddard Institute for Space Science, the first prominent scientist to openly warn America about global warming. When Hansen told a U.S. Senate panel in 1988 that climate change was already transforming our world, he was dismissed by many as an alarmist and a kook. Yet nearly twenty years of scientific inquiry and explicit physical evidence gathered worldwide have shown him to have been absolutely, unerringly correct all along.

So when Hansen and other leading climate scientists point to Greenland as the potential source of truly catastrophic sea-level rise, one tends to pay attention. The Greenland ice sheet is incomprehensibly huge, the second largest in the world, containing 900,000 cubic miles of frozen water. It, too, is melting, of course, shrinking at a rate of 21.6 cubic miles annually since 1993. That's enough to cover my state of Maryland with nine feet of water every year. This is land-based icemelt, so it contributes directly to sea-level rise, unlike the floating sea ice of the arctic.

The IPCC's projection of one to three feet of sea-level rise takes into account this melting ice but assumes that the Greenland ice sheet will remain more or less intact, melting peripherally but not collapsing entirely and sliding into the ocean. This is good news, since a complete meltdown would raise sea levels not three feet or ten feet but a disastrous *twenty-three* feet! Under this scenario, much of south Florida would simply disappear and the Jefferson Memorial would be under thirteen feet of water.

But three recent discoveries now cast doubt on the IPCC's assumptions about Greenland. First, in 2005, scientists found that the ice sheet was melting twice as fast as measurements showed just five years earlier. This rapid acceleration meant that enough Greenland meltwater was now entering the ocean every twelve months as to rival all the water in Lake Erie.

Second, scientists have found that some of the surface icemelt in Greenland is actually filtering down through crevasses to the bedrock upon which the entire ice sheet rests. This water then acts as a kind of lubricant that could be ushering the entire body of ice toward the sea as if on a greased rail. Whether this process has accelerated the ice sheet's disappearance during past epochs of climate change is unknown. What *is* known is that the ice sheet *has* disappeared before.

Which leads to a third recent discovery about Greenland. Evidence gleaned by climate historians from ice cores, ancient coral reefs, and other natural climate records now suggests that most of the Greenland ice sheet has melted away in the past whenever the earth's temperature has been about four to six degrees warmer than it is right now. Just four to six degrees. The evidence also suggests that the West Antarctic ice sheet, the world's second-largest land-based piece of ice, would disappear under this scenario as would part of the East Antarctic ice sheet. These three ice sheets combined would create a kind of icemelt tidal wave, causing sea-level rise of a full *eighty feet* worldwide.

There's no point in even imagining what that world would look like. It's simply unthinkable. Which is why NASA's Hansen has raised his voice even louder in recent years, imploring world leaders to imme-diately embrace the feasible clean-energy technology and efficiency gains that would keep the planetary warming below two degrees this century.

Otherwise, quite frankly, no matter how well we armor our coast-lines and build massive floodgates and construct weird "amphibious" houses that literally float during floods, there will be no "accommodat-ing" sea-level rise. Nor will there be an orderly, manageable "planned re-treat." The retreat, instead, will be messy and fraught with surprises. It will be unplanned and full of pain. As a nation we now have abundant, firsthand evidence out of Louisiana to support this sobering prediction.

When I stood in the Marian Koshland Museum in downtown Washington, D.C., and pushed the flood button for the Chesapeake Bay, I had an overwhelming sense of déjà vu because, six years earlier, long before Katrina hit, I had pushed a similar flood button for coastal Louisiana. It was on a website run by a Louisiana conservation group, and it showed not what would happen from three feet of water rising over the *next* hundred years. It showed what had *already* happened to the Louisiana coast over the *last* hundred years from three feet of water. You clicked a button and saw the coastal maps from 1900 and 1950 and 2000. You saw the islands and wetlands vanish and the water pour closer and closer to New Orleans. This map demonstrated that the Chesapeake Bay scenario was no fantasy. What was coming to one place had already come to another.

New Orleans, like man-made global warming, was a sort of unsu-pervised, large-scale, and very dangerous experiment. Let's throw three feet of water at a low-lying U.S. coastal city and ignore the warnings till the bitter end and then see what happens.

The results are now in, of course. Nearly thirteen hundred people died in Louisiana and the economic loss exceeds $200 billion—not be-cause of eighty feet of sea-level rise or twenty-three feet, but because the ocean rose just three feet higher relative to the land. Over a million Louisianians fled as well. They fled in the most wrenching and disorderly way possible, all of them "displaced" in an explicit act of mass re- treat. Wherever these people may have gone, two things are certain: Al- most all of them have relocated somewhere farther away from the coast, and a huge number of them will never return. The ocean bullied everyone back a bit. That's the power it has. And the retreat, though unplanned, is pretty much permanent.

6

Killer Hurricanes: Exporting Katrina to the World

IT'S NOT ENOUGH, it seems, that the seas are get-ting bigger, heading toward a possible three-foot rise in the coming decades. And it's not enough that hurricanes on their own, without human enhancement, are powerful enough to destroy entire coastlines. Now comes disturbing evidence that the same warmth that's making the ice melt and the water rise is also cooking hurricanes into monstrous storms endowed with longer lives and lots more power.

Whether one hurricane in particular, Katrina, was made more intense by global warming is impossible to say. What *is* certain is that two things worked together to kill New Orleans: a relative sea-level rise of three feet followed by a gigantic storm. Now every coastline in the world faces similar sea-level rise at the same time that the largest, most powerful hurricanes, according to piles of new scientific data, are becoming more frequent. This, of course, makes New Orleans an even more compelling case study of future conditions. To visit the city in person, as I did ten weeks after Katrina, is to assume the role of a time traveler, in fact.

Let me first say that I've experienced my full share of bad things in this world. As a Peace Corps Volunteer in poorest Africa, I lived among children wasting away from chronic malnutrition, some dying right before my eyes. As a journalist, I've covered armed conflict in Latin America and bloody "crack wars" in Washington, D.C., and stories of environmental ruin on virtually every continent. And on a personal level I've suffered my allotment of illness, bad luck, betrayal, and the death of loved ones.

But I can say with complete honesty that the darkest, most disturbing, most emotionally unsettling thing I've ever experienced in my life is my visit to New Orleans, Louisiana, ten weeks after Hurricane Katrina. For anyone who's not been to the city since the calamity, the newspaper articles you've read convey a lot about what happened. So do the radio reports and TV coverage.

But what you can't even begin to appreciate until you go there— what I, myself, was unprepared to take in—is the enormous *scale* of it all. The vastness of the annihilation and loss—from the empty skyscrapers to the children's toys hanging from trees to the ocean of uninterrupted darkness that still washes over much of the city at night— this cannot be captured secondhand. To wear out a pair of shoes or burn up a tank of gas and still not see one wholly functioning neighborhood or one square block of intact businesses is to experience something no one now alive has ever known before, much less witnessed firsthand, on American soil. It is another world entirely, New Orleans, and to go there is to experience grieving beyond words.

And, in my case, beyond limits. For to visit New Orleans with the simultaneous knowledge that all that ruinous saltwater, now safely returned to the sea, is simply biding its time, getting warmer each year, steadily planning a comeback, is to know a deeper sadness still. To visit with the knowledge that the sky itself is in the grips of a fever, storing up power for even darker events to come, is to see and feel amid the

muddy, wrecked streets of New Orleans a strange new universe looming just beyond America's near horizon.

To SAY THE 2005 HURRICANE SEASON was unusually "active" or
"powerful" is like saying a nuclear explosion is "active" compared to a
conventional TNT bomb. The truth is there's no adequate way to describe what happened during that 2005 storm season. One thinks of
the Inupiat people of Alaska, faced with rapid Arctic warming, who
have no word in their language for the barn owls and wasps arriving in
recent years because they've simply never seen such creatures before.
The 2005 hurricane season shattered so many records and brought
such weird and unprecedented storm behavior, that the season as a
whole was simply unrecognizable compared to all previous storm
years.

It began, this season, with Tropical Storm Arlene, forming unusually early—June 9—in the lower Caribbean without inflicting meaningful harm. It ended seven long months later when Hurricane Zeta
took shape on December 30, a full month after the official close of
the Atlantic hurricane season. Remarkably, Zeta kept churning 2,600
miles off the coast of Florida even beyond January 1, 2006. She finally
dissipated January 6, the first storm ever to survive so long into the
next calendar year. The 2005 season was so "active" it didn't even end in
2005.

In between Arlene and Zeta came a veritable traffic pile-up of
records. Never before in the Atlantic Basin (Gulf of Mexico, Caribbean Sea, and the Atlantic) had twenty-seven named storms formed in
one season. The old record was twenty-one, set in 1933. Never before
had fourteen full-blown hurricanes formed in a single season (old
record: twelve in 1969). Never before had four major hurricanes hit the
United States (old record: three in 2004). Never before had three Category 5 hurricanes formed in a single season (old record: two in 1960

and 1961). Never before had seven tropical storms formed before August 1 (old record: five in 1977).

Like Zeta, many of the storms themselves had extremely odd characteristics. In October, Hurricane Vince formed west of Spain near the Madeira Islands in the northeastern Atlantic. Very, very few hurricanes ever form so close to Europe. Then Vince actually struck the continent, coming ashore as a tropical storm in Spain, a country that had never experienced such an event going all the way back to its ancient explorers, who reported such violent storms only in the far-off New World.

(A year earlier a similar oddity occurred when the first-ever hurricane formed in the South Atlantic, hitting Brazil with ninety-mile-per-hour winds and killing twelve people. The Brazilian weather service, understandably devoid of any naming system for storms that never existed before, was unsure what to call it, finally deciding on Catarina, after the state where it made landfall.)

Not to be outdone, late October 2005 produced Hurricane Alpha, forming in the Caribbean near the island of Hispaniola. Alpha is famous simply for her name. Never before had meteorologists run through the entire list of twenty-one preassigned names reserved for a single Atlantic storm season. But when Alpha formed, representing the record-setting twenty-second storm of the 2005 season, meteorologists had to turn to the Greek alphabet. Before the year was out, the world had watched storms Alpha, Beta, Gamma, Delta, Epsilon, and Zeta rage across the Atlantic and Caribbean.

But 2005 will be remembered most for the sheer destructive force of the hurricanes it spawned. If you make a list of the six most powerful hurricanes *ever* recorded in the Atlantic Basin, you find that three of them occurred within the eyeblink of a 52-day period in 2005. Those storms are Katrina, Rita, and Wilma. That fact bears repeating. It took 153 years of human record keeping to identify one half of that A list of top-six hurricanes. It took 52 days to record the other half.

Katrina became a Category 5 storm on August 28 in the Gulf of Mexico with wind gusts up to 175 miles per hour and a barometric pressure of 26.64 inches, the sixth lowest ever measured. She came ashore as a high Category 3 with a record surge of 29 feet measured at Bay St. Louis, Mississippi. Twenty-three days later, Rita set the mark as the most powerful hurricane ever recorded in the Gulf of Mexico and, with a barometric pressure of 26.49 inches, became the fourth most poweful ever in the Atlantic Basin. She had sustained winds of 175 miles per hour and gusts as high as 235, and her enormous swirl of clouds was, at one point, nearly as big as the Gulf itself.

Barely three weeks after Rita, Hurricane Wilma formed off the coast of Jamaica and seemed to live on forever, hitting Cozumel, Mexico, then Playa del Carmen, then Cancun, then Key West, Florida, then mainland Florida, then the Bahamas. In the wee morning hours of October 22, NOAA "hurricane hunters" flying at over forty thousand feet in a twin-engine Gulfstream IV jet—named Gonzo for its unusually shaped nose—dropped twenty-three measuring probes into Wilma's path and were stunned to read sustained winds *above* 175 mph and the lowest barometric pressure ever recorded in the Atlantic Basin: 26.05 inches. This was the queen of all Atlantic storms, the undisputed champion.

These extreme physical characteristics—record winds and low pressure—translated directly into record impacts on human beings, of course. When three of the biggest monsters the world has ever known form within fifty-two days of one another, there's plenty of pain to go around.

Way back in 1992, when Hurricane Andrew struck south Florida killing twenty-three people and inflicting a staggering $43.7 billion in economic losses, the impact was so great it bankrupted no fewer than eleven Florida insurance companies. For years afterward, Andrew was held up as the nightmare, once-in-a-century storm. It just couldn't get much worse than Andrew.

So what, then, do we call Katrina? The millennium storm? The economic loss from Katrina is more than *four times* what Andrew inflicted. The human loss is fifty-six times greater than Andrew.

Rita, meanwhile, inflicted "only" $6 billion in damage because she had the good manners to come ashore along one of the least populated stretches of the entire Gulf coast. She still atomized coastal communities like Holly Beach and Cameron, Louisiana, and destroyed or damaged 70 percent of the housing stock as far inland as Lake Charles. The salinity of in-rushing ocean water, meanwhile, contaminated vast rice fields and cattle fields all along southwest Louisiana, not unlike the ancient fields of Carthage sown with salt by invading Roman armies.

And then there's Wilma, referred to in one NASA report as Katrina's "evil twin." Wilma struck the west coast of Florida just below Naples with such force that even after crossing the Florida peninsula she had enough power to shatter much of south Florida's *east* coast. The storm cut off power to nearly six million people, caused $10 billion in damages, and was the worst hurricane to hit Fort Lauderdale in fifty-five years. And Wilma came from the *west*. She came over land, through the back door.

The Insurance Information Institute, a New York–based organization funded by the insurance industry, now ranks Wilma as the sixth-costliest hurricane on record, with Katrina all alone at the top, of course, with damages in the stratosphere. More amazing, the Institute now reports that seven of the ten biggest loss-creating hurricanes in history occurred in just two years: 2004 and 2005.

Anyone older than kindergarten age can spot a trend here. But what exactly *is* happening? Why so much sudden hurricane fury and economic loss? Why the skyrocketing death rate? Hurricane Stan actually surpassed even Katrina in the fatality category, burying two thousand Guatemalans in mud in October 2005.

And why did Max Mayfield, head of the National Hurricane Center, tell a Florida TV station at the start of the new year that present trends suggest the 2006 hurricane season could make 2005 look "mild." With dead seriousness, he then advised all coastal residents who don't have the means or inclination to evacuate quickly to keep an ax in their attic so they can bust through the roof if the surge tide finds them in 2006.

But are these storm seasons just particularly powerful years in an otherwise natural cycle of hurricane variability? Or has something changed? Is this, dare we even think it, the "new normal"?

DESPITE PRECISE and rapidly growing scientific observations world-wide of a changing climate over the past twenty-five years, hurricanes have seemed to show a strange and stubborn reluctance to change in any way. A general consensus has remained strong among tropical cyclone scientists that there is no evidence of unnatural hurricane patterns in the Atlantic over the last century. Hurricane activity continued to alternate normally between multidecade phases of high activity and similarly extended phases of low activity. This cycle was driven by a host of factors ranging from changes in ocean currents and salinity to altered rainfall patterns in the Sahel region of Africa. And even as glaciers melted worldwide and entire ecosystems were turned on their heads, the great warming appeared to change hurricanes not at all.

This seemed odd for several reasons. First, almost all the computer models used by scientists to estimate future impacts of global warming predicted more intense hurricanes. The atmosphere was warming, which was heating the oceans, and hurricanes get most of their power from warm water. One summary of twelve hundred computer simulations published in the *Journal of Climate* in 2004 revealed that soaring carbon dioxide levels could triple the number of Category 5 hurricanes in coming years.

But as late as 2005 there was still no convincing evidence that this was already beginning to happen. If everything in nature is connected to everything, how could the planet's largest, most complex storms remain unaltered amid massive transformations all around?

Ironically, it was right in the middle of the tumultuous 2005 storm season that two new studies hit the scientific community with the force of hurricanes themselves. The old consensus was only partly correct, the studies revealed. There has been no change worldwide in the *frequency* of tropical cyclones (the catchall name for low-pressure systems over tropical waters) in recent decades. That much was still clear. When the number of hurricanes in the North Atlantic increases one year, for example, the number of cyclones in the North Pacific, on the other side of the planet, tends to decrease correspondingly and vice versa. Over years and decades, these various trends tend to cancel each other out, creating a global equilibrium in the number of storms.

But frequency is not the same as *intensity*. And by studying data from over four thousand hurricanes over the past fifty years, the 2005 studies presented empirical evidence for the first time showing that, yes, the storms have been gaining in power all along, fueled by warmer water and warmer air.

Kerry Emanuel, a well-known storm guru at the Massachusetts Institute of Technology (MIT), was the first to step forward. Skeptical for years of any connection between planetary warming and storm intensity, he telegraphed his change of mind in July 2005. He abruptly withdrew as co-author of a scientific paper reflecting the old consensus view that hurricanes were showing no significant changes. According to the *Washington Post*, Emanuel e-mailed a co-author, saying, "The problem for me is that I cannot sign on to a paper which makes statements I no longer believe are true. I see a large global warming signal in hurricanes."

The very next month, Emanuel published a paper in the prestigious British science magazine *Nature* showing that the average power of storms in both the Atlantic and the North Pacific had *doubled* in the past thirty years. He determined this by using a special method for measuring the storms' wind speed and life span. Significantly, the growing intensity of storms corresponded closely with rising sea surface temperatures worldwide.

Then in September, while Louisiana was absorbing the brunt of Katrina and Rita, a similar study emerged from a team of researchers that included noted meteorologist Greg Holland of the National Center for Atmospheric Research in Boulder, Colorado. This study, published in the journal *Science*, showed that the number of Category 1, 2, and 3 storms worldwide has fallen slightly in the past thirty-five years, while the number of Category 4 and 5 storms—the most powerful ones—has grown tremendously. In the decade of the 1970s, an average of ten storms occurred annually in the Category 4–5 range. But from 1990 forward, the yearly average has almost doubled to eighteen. "We're talking about a very large change," Holland told *Time* magazine.

Again, Emanuel's data reveals that hurricane intensity is growing in tandem with rising sea surface temperatures. The meteorological explanation for this is pretty straightforward. Hurricanes are monstrous systems that feed ravenously on both heat and moisture. These storms form only over water that is at least 80 degrees Fahrenheit to a depth of 150 feet. The air above a warm ocean, in turn, holds more moisture. As sea-surface temperatures increase, so do the moisture levels in the air above, and this moisture constitutes the raw fuel for hurricane growth.

A storm gets its start when an area of low pressure forms over warm seawater at least three hundred miles north of the equator. The low pressure draws in air from surrounding areas of higher pressure, and the earth's rotation causes those winds to swirl counterclockwise in the Northern Hemisphere. At the center of the storm, warm air already rich with water vapor from the ocean heat begins to rise up rapidly. This ascending air magnifies the suction effect, drawing in more surrounding air. Eventually, much of the moisture in this rising air condenses to rain and falls back to earth, a process which itself releases a tremendous amount of latent heat. This heat triggers even stronger updrafts of air, which in turn evaporates more water from the ocean surface, creating a self-reinforcing vortex of swirling clouds, rain, and wind. If strong upper-level winds don't disrupt this cycle, the storm becomes a hurricane once the spiraling air speed reaches seventy-four miles per hour.

Again, warm water and moist air are the driving forces. And with global warming, both of these elements are in greater supply worldwide. The mean surface temperature of the world's oceans rose a full one degree in the twentieth century in almost perfect parallel with the rising atmospheric temperature. Most of that warming has occurred in the later half of the century. And sure enough, satellite measurements since 1988 have recorded a full 4 percent rise in water vapor over the world's oceans, supplying both extra moisture and heat power for any tropical storm that comes along.

Suddenly, the seemingly endless hurricane records of 2005 start to make frightening sense. By NASA's measurements, 2005 was the warmest year ever recorded. The year also racked up more hurricane deaths and destruction than the previous ten storm seasons *combined*. The words of MIT's Emanuel—"I see a large global warming signal in hurricanes"—are now on the lips of tropical storm meteorologists worldwide.

People along the U.S. Gulf coast are also right to point out that, intensity aside, the sheer number of hurricanes has grown in recent years compared to past decades. This, however, appears to be part of a natural cycle in the Atlantic Basin of inactive hurricane phases followed by active phases. For example, the late 1920s to 1969 saw great hurricane activity year after year only to be followed by extraordinarily low annual numbers from 1970 to 1994. Then, in 1995, the Atlantic Basin entered the active phase it's still in now, and almost all storm scientists expect this normal upswing in numbers to last another decade or two.

Again, what isn't normal, according to the Emanuel and Holland studies, is the rising number of gigantic storms. When Hurricane Katrina crossed the Florida peninsula on August 25, she was a Category 1 storm. Four days later she slammed into the Louisiana coast as a high Category 3 after first spending time out at sea as a whopping Category 5. What explains the big power surge between Florida and Louisiana? Super warm water, that's what. The storm passed over surface temperatures in the Gulf that were up to 4 degrees Fahrenheit higher than normal. Similarly, high temperatures gave the same boosts to Rita and Wilma soon after.

These 2005 storms came right on the heels of the appalling storm season of 2004, of course. After the quartet of Francis, Jeanne, Charley, and Ivan ransacked Florida in quick succession in 2004, setting records for damage, many Americans came to appreciate the original Caribbean Indian meaning of the word *hurricane* as "evil spirit and big wind."

Finally, as if the news weren't bad enough, a less-publicized aspect of Emanuel's study reveals that these "big winds" are also blowing *longer*. His data confirms that the cumulative annual duration of storms in the North Atlantic and much of the North Pacific has rocketed up by 60 percent in recent decades.

The storms, in short, just stick around longer while adding more and more muscle with each season. Imagine for a moment the combined electrical production of every power plant on earth running flat out for six full months. Imagine every nuclear facility, every wind farm, every coal-fired power plant in every country, all put together in this way. That's the power generated by a single average hurricane. But Katrina, Rita, and Wilma were each at least ten *times* more powerful than even this.

If this is what we can expect year after year in a hot, future world, then an evil wind is blowing with great gusto indeed.

ONE HUNDRED AND SIXTY-SEVEN HURRICANES struck the United States during the twentieth century. Assuming a repeat performance this century, that's 167 greatly enhanced opportunities for storms to cause human harm courtesy of a changing climate. All 167 storms will find coastlines that have surrendered up to three full feet of elevation to the sea, creating ideal high-tide conditions no matter where the storm strikes or when. Indeed, just one foot of sea-level rise, according to FEMA, would increase coastal flood damage between 36 and 58 percent, even without factoring in a growth in hurricane intensity.

But there *will* be growth. In fact, if Holland's study is any guide, more than a third of those 167 storms will be Category 4 or 5 monstrosities, nearly twice the figure for the middle twentieth century. No section of the U.S. Gulf coast or Atlantic coast is safe from this heightened threat, of course. Whether you live in a Corpus Christi beach bungalow or a high-rise condo along the Jersey shore, this is, like it or not, your reality.

But some areas are more vulnerable than others. FEMA, in fact, points to three cities at the top of its list of potential hurricane megadisasters. These are New Orleans, Miami, and New York City. Each city deserves a separate examination here in light of accelerating climate impacts.

No further proof is needed that New Orleans is vulnerable to hurricanes, of course. The sky has already fallen. The results are in. The bad news as the city rebuilds is that, even if it gets the full $14 billion needed to reconstruct coastal wetlands and barrier islands, New Orleans is doomed to Atlantis status if global warming brings all that additional sea-level rise followed by more ferocious hurricanes. Even the great land-building capacity of the Mississippi River can't keep up with that much new Gulf water magnified by the type of wind and surge damage that comes with Category 4s and 5s.

To survive into the twenty-first century, New Orleans not only needs all of America to help it "reengineer" its fragile coastline and rebuild its fragile levees, it needs all the world to help slow down global warming. New Orleans, in short, needs a heck of a lot.

As for Miami, you need to know only two things in the context of global warming: (1) much of the city is barely three feet above sea level right now, and (2) more than *one-third* of those 167 hurricanes that struck America during the last century landed on Florida. Miami, and all of south Florida, is right in the middle of a great hurricane romper room. The city is always due for another strike. *Always.*

Meanwhile, the ocean water is rising quickly all around Miami. To the east, already, nothing protects the city from a head-on hurricane strike. To the south, as mentioned earlier, the buffering network of keys and coral reefs may disappear completely. And to the west, pretty much all the land is made up of current or former swamps and marshland, part of the vast Everglades ecosystem, two million acres stretching from the Keys in the south to Lake Okeechobee in the north to Naples in the west. This wetlands complex constitutes much of the lower third of the entire state of Florida and most of it is just a few feet above sea level now. Add three feet of ocean rise and some estimates have the entire lower Everglades turning to open water, right on Miami's backside. The Gulf of Mexico will just pour in, dramatically changing the very shape of the state of Florida.

Hurricane Wilma struck Florida south of Naples, crossed over the Everglades, and still survived as the worst hurricane Fort Lauderdale had seen in fifty-five years. Imagine if Wilma had crossed over not a hundred miles of marsh grass and mangroves but a vast, hot expanse of open water on her way to Florida's densely populated east coast. The outcome certainly would have been much, much worse.

Florida's vast population is another source of trouble. The state has doubled in population to 15 million just since 1980. Indeed, just two south Florida counties—Broward and Dade—have more people today than lived in all 109 coastal U.S. counties from Texas to Virginia in 1930. There are just a lot more people and buildings for a hurricane to destroy today along most of the U.S. seacoast.

Of course, many observers point to this explosive growth in coastal development as a major reason overall hurricane damage has risen so sharply in recent years, and there's some truth to the claim. But we now know that hurricane wind speeds *have* increased, nearly doubling in the past fifty years. And as wind speed grows, the pressure it puts against any object in its way grows at a disproportionate rate. A 25-mph wind, for instance, creates roughly 1.6 pounds of pressure per square inch. That force rockets to 450 pounds in 75-mph winds, and nearly triples to 1,250 pounds in 125-mph winds. No wonder property damage tends to increase tenfold each time a hurricane moves one category higher in the standard Saffir-Simpson intensity scale. A Category 4 storm, for example, with wind speeds in excess of 148 mph, will produce on average up to 250 times the damage of a Category 1 storm with winds of 74 mph, according to NOAA.

So, yeah, more people are now in the way of storms than in the past. But more *storm* is in the way of people, too. It's no wonder then that Munich Re, the world's largest insurer of insurance companies, asserts that extreme weather events related to global warming have contributed to a rise in worldwide insurance losses over the last fifty years. Insured losses in 2005 alone exceeded $83 billion, the highest total ever.

AND THEN THERE'S NEW YORK CITY, the great sleeping giant of hurricane disaster sites. So many mutually reinforcing factors point to catastrophe in America's largest city that, in many ways, it's even more frightening than New Orleans. "The Bobbing Apple" might be the name we use once the perfect storm arrives here.

Few hurricanes strike land this far north, of course, most drifting harmlessly out into the upper Atlantic, pushed there by strong westerly winds. But every forty to seventy years, a major storm does slam into the New York City region. The great hurricane of 1821 passed right over Manhattan and basically cut the city in two, with the Hudson and East Rivers merging all the way up to Canal Street. At the Battery, shocked city dwellers watched as water rose as fast as thirteen feet in one hour. Saving the city from total annihilation was the storm's lucky arrival, like Gloria, at low tide.

Another major storm struck in 1892, then another in 1938 when the borderline Category 4 "Long Island Express" passed through the outskirts of greater New York, inflicting widespread death and destruction across New York state, New Jersey, and much of New England. But that storm, sixty-eight years ago, was the last major hurricane (Category 3 or above) to strike the New York Metropolitan region. It's now a matter of when, not if, a big hurricane will strike again, according to meteorologists. And history says "when" is very soon.

The basic geography of New York City makes it a worst-case landing strip for any major hurricane. The city sits at the vertex of a giant right angle created by the land platform of Long Island and the north-ern shore of New Jersey. In the middle, looking like a huge catch basin, is New York Harbor, the gateway to the city. In the face of a major storm surge, the harbor acts as a funnel. The wall of seawater will plow into the harbor and then roar up the Hudson and East Rivers, where it will become suddenly trapped. It'll get backed into a corner with nowhere to go. Nowhere, that is, except up. In an eyeblink, the runways of La Guardia and JFK Airports will be under eighteen feet or more of water. Compounding matters is the fact that New York sits on an ex-tremely shallow continental shelf, which causes any surge to pile up on itself even before it reaches the city. These factors together give Gotham some of the highest storm-surge values in the United States.

It's not just the city's surface that will flood from a Category 3 storm, with almost all of lower Manhattan south of Broome Street under up to thirty *feet* of water. It's also the subsurface. Much of New York, like New Orleans, is below sea level. It's actually *underground*, in fact, in the form of subway tracks, car tunnels, multistory parking garages, basements, and utility tunnels. And in a big storm, it all floods. The city's surge maps show that the Holland and Brooklyn Battery Tunnels will become totally filled with seawater in a Category 3 storm.

An unsettling preview of things to come in New York arrived in December 1992, when a powerful nor'easter struck the city. It raised sea level at the southern tip of Manhattan by eight feet, flooded the Brooklyn Battery Tunnel with six feet of water, forced La Guardia Air-port to close, and shorted out the entire New York subway system, stranding passengers on trains and in stations (saltwater conducts elec-tricity, causing shorting, and it's corrosive). The whole city was com-pletely shut down, paralyzing the lives of 21 million people in the tristate metro area—7 percent of the U.S. population. And this wasn't even a hurricane.

What's more, none of the hurricane forecasts described here takes into account the prehurricane "flood" that global warming will roll into the city. Amplified by local geological conditions, one U.S. government study predicts that up to 2 feet of ocean rise may occur by 2050 in New York and 3.5 feet by 2080. The next Big One could find the city already totally saturated with water around the edges, hanging on by a thread.

Finally, the same warming that raises sea level will also likely endow New York's next Big One with much stronger winds. These winds will push still more surge water into the city. The wind itself will then blow through the city at, say, 120 miles per hour, with impacts scarcely comprehensible. Many of the city's two million trees, their roots wrapped tightly around all manner of buried telephone and electrical wires, will fall.

And that's just at street level. A hurricane's winds increase in force as you go up, in some cases nearly doubling in speed at 350 feet. This would deliver a super-enhanced body blow to every building in the city above thirty stories, which is to say pretty much the entire skyscraper forest of Manhattan. Who can even imagine the scene of flying glass and dislodged masonry filling the air and falling to the street everywhere from Midtown to the Financial District?

Whatever hurricane protection might come from the city's proposed floodgate barriers, one thing's for sure: No barrier will stop the surge of wind power above the ground. And those floodgates themselves better be built on a colossal scale to handle the kind of water that's bound to come unless the United States joins the rest of the world *right now* in slowing down the whole warming process.

One wonders, however, if it'll take a global warming–enhanced, long-overdue, direct-hit hurricane in New York City before the nation commits to building the massive floodgates at all. Perhaps then we'll consider a national switch to renewable energy, too.

But if that's what it'll take, we should prepare ourselves for an economic wound and humanitarian crisis that outstrips all the images and impacts of September 11 and Katrina combined. One 1995 study, co-authored by the Army Corps of Engineers, envisions a plausible scenario where untold thousands of New Yorkers—fleeing severe wind and rain and a hail of falling glass and flying bricks crashing through office windows—rush in a panic to the shelter of subway tunnels. And that's where they all die, drowned by the hurricane's surge waters that rapidly fill the tunnels and stations.

And in a page straight out of the emergency preparedness fiasco of New Orleans, the New York City Office of Emergency Management will be hard-pressed to help out. Its main office in Brooklyn will itself be under several feet of water.

IF ONLY IT WERE ALL A DREAM: oceans warming, hurricanes growing, Greenland melting, Florida drowning. If only it were pure science fiction when we talk about towering floodgates and cyclones hitting Europe and sea-level rise of three feet or twenty-three feet or eighty feet.

But these are the enduring "science facts" of our world unless we seek a better option soon. The more you look at those facts, the more you realize that the two coastal response strategies—adaptation and retreat—just won't hold. If by "adapt" we mean staying put along our shorelines while maintaining some semblance of what we've come to call modern life, then the growing chaos and expense and risk now forecast for us seems far too great to inspire any degree of confidence. Why would we want to adapt to such a place anyway?

As for retreat, where are 90 million Atlantic and Gulf coast residents supposed to go? The stable economy and abundant food supply of America in 2005 allowed two million Katrina evacuees (Louisiana, Mississippi, and Alabama refugees combined) to step back from the ocean and more or less land safely elsewhere across a bountiful nation. But nothing happens in isolation on this planet. In coming decades, as global warming batters our coastlines, the nation's interior will not magically carry on as a welcoming, unchanged land of milk and honey. With a mid-level warming of four to five degrees on the IPCC projection range, Kansas could become a scrub desert. What in the world will we eat if our national breadbasket is a scrub desert? And three-quarters of all the water consumed in the western United States begins as melt-water from mountain snowpacks that are rapidly disappearing because of heat and drought. What will we drink? And with record heat-wave deaths in our Midwestern cities and the possible widespread advance of mosquito-borne diseases, how will we cope?

As these future stressors pile up, running nonstop on so many different levels, even one Katrina disaster might be impossible for the nation to absorb in 2015 or 2025. Never mind two or three Katrinas every summer. I can't help but wonder what I will do in my own region. The Chesapeake Bay, warmed and greatly enlarged by sea-level rise, will be an enhanced landing strip for newly pumped-up Atlantic hurricanes. A storm like Wilma, with an unusually tight and compact eye wall, will be able to travel in the future to within a few miles of the Washington Beltway and my house, refueled all along the way by the hot and wide-open water of a bay no longer buffered by vast marshes and islands. Where will I go after fatigue sets in from the first rebuilding and then the second and third?

Should I head for the nearby Appalachian Mountains where my friend Mike Tabor, the farmer, is already thinking of quitting the land because the weather's so violent and unpredictable? Do I head to my native Georgia where the chaos of drought and heat will likely convert vast Southeastern forests to grassland and where, certainly, I'll compete with millions of other "evacuees" for work and social services?

In truth, I don't plan to pursue either of these options. That's because I don't see them as options at all. I reject them for myself and my nine-year-old son and for all future generations of Americans. I reject them because they are not the only choices before us. As you'll see in the following chapters, there's another way, a clear-eyed, affordable, and feasible way to fight back against the warming.

So instead of contemplating a distant retreat to the mountains of western Maryland or the suburbs of northern Georgia, I decided in November 2005 to visit the sad and broken landscape of New Orleans, Louisiana. I went to the city so soon after Katrina with the hope that the images I'm about to describe will help convince you, too, to reject the trauma of future retreat and the false hope of adaptation. I went both to mourn a city I love and to gain a sneak preview of a warming world we can still avoid. Faraway glaciers and thermal expansion and buttons on a museum wall are, in the end, mere abstractions no matter how hard we try to make them real in our minds. What does a major U.S. city actually look like and feel like and sound like when a mammoth hurricane joins forces with three feet and a hundred years of relative sea-level rise to create a culminating, catastrophic weather event the likes of which America has never seen before? You have to go there to comprehend it.

So I bought a plane ticket to New Orleans and traveled, as if by time capsule, to a future world I hope will never come even as the sound of its rumbling surge tide lets us know it's rapidly pouring in from afar.

THE SENSORY SHOCK WAVES began immediately that November day, 2005, as my plane approached New Orleans from the east. There below my window I see a city awash in a bright blue ocean. It stretches from the bungalows of suburbia to the downtown skyscrapers to the airport terminal. It's not water, of course. The water's all gone now. It's plastic: countless blue sheets of rooftop tarpaulin. These so-called FEMA tarps—tens of thousands of them—were handed out by the agency right after the storm and were meant to be used for thirty days or less.

But two and a half months later, they still cover, by my aerial estimate, a full quarter of the city. The scale gives the impression almost of decoration, a kind of Christo creation large beyond imagining. New Orleans, what's left of it, is a shiny, blue tarp city, artful in its death pose, huddled under plastic. Unhindered by old landforms, Katrina got inland with enough wind power to shave the tops off tens of thousands of buildings.

Within minutes I'm on the ground and driving through the heart of the city on Interstate 10. Here, on this elevated highway, is where so many of the hungry and sick lay stranded, above the water, for days after the storm. But today there's traffic—cars, pickup trucks, eighteen-wheelers—which seems hopeful until I notice the license plates: Texas, Florida, Georgia, Arizona. The drivers are simply passing through. There's nothing to visit here. It's a giant accident scene, this town, where people slow down to gawk a bit and then move on.

I, too, stare down from the elevated highway, peering into a ghostly gray urban landscape that looks like it's been hit by a neutron bomb. The buildings are still standing but the people are all gone. They've vanished. No people.

What does exist are huge piles of debris across every sidewalk, and traffic lights that aren't working, and dark, straight, horizontal lines on every wall and every tree where the water peaked and left a bathtub-ring stain five or ten or fifteen feet off the ground.

And there are cars. Dead cars. Everywhere. Hanging sideways in trees. Lying upside down on lawns. Joined in weirdly suggestive pairs as if copulating. Nearly *a quarter million* cars litter the town, flooded out, covered with filth, abandoned.

I exit first into the area called New Orleans East, in the Chalmette community of hard-hit St. Bernard Parish. A National Guardsman in combat fatigues waves me through a checkpoint and suddenly I've arrived, ground level.

Who can comprehend water rising as fast as a foot per hour? Even those who were there, who survived, struggle with words. Down the streets the water came, smashing through bedroom windows, breaking down doors. My car now rolls through an endless graveyard of houses, each a molding, tilting headstone marking a family swept away to Houston or Atlanta or Baton Rouge after the levees failed and the tidal wave came in.

There are lots of blue tarps here, too, but from the ground the sense of surreal artwork has changed. The angles and lines are so distorted, everything out of proportion to the normal, that I now feel as if I'm driving through a Salvador Dalí painting. A shattered bass boat is on a roof, dripping clumps of marsh grass and muck onto a piano that's protruding from a melting, soggy, collapsing living room wall. A stuffed Winnie the Pooh, as big as a child, has landed atop a backyard barbecue pit, disintegrating above the cracked bricks. A telephone pole is upside down, its base having snapped and flipped skyward while the wires never broke.

Maybe one in ten houses is being salvaged here, gutted by a strange combination of Mexican immigrants and National Guardsmen. I see a group of Guardsmen pull over in a camo personnel carrier, a wheelbarrow tied to the radiator grill. They pull on protective white hazmat jumpsuits while the commander tells me he's just back from Iraq. His job now is to remove destroyed "white goods"—refrigerators, washing machines—from flooded homes.

"Iraq is nothing," he says. "Here it's like Beirut or Berlin, Germany, after the war or something. It's that total."

A block away I enter a house being gutted by Mexican workers in masks. Two inches of mud coat the floor. I have a headache within minutes from the suffocating mold covering the walls and ceiling. A limping worker—he stepped on a nail yesterday—points out there's no water line in this neighborhood. No line. It rose that high. The men pile everything onto the curb—destroyed carpeting, lamps, bedding, underwear, Sheetrock—leaving only wall studs.

These scenes repeat themselves again and again for hours as I drive through Chalmette and then into New Orleans's Ninth Ward and then the Lakeview area north of downtown. Major streets are lined with endless roadside signs—about the size and number you'd see in a hot political race. But these signs are advertising disaster services: "House Gutting Real Cheap," "Mold treatment/roofing," "Furniture Removal and Sheetrock Tear Out."

My question of where all this debris is going is answered when I turn a corner onto Westend Boulevard. There, right in the middle of the city, measuring a block wide and many blocks long and three stories high, lie the exposed and rotting entrails of the city of New Orleans.

The Guardsmen and Mexicans and shopkeepers and homeowners all bring their debris here, dumping it onto the sprawling, wide-open median of this once graceful boulevard.

I tilt my head back and look up at the immeasurable tons of broken dishes and rotting blue jeans and damp carpeting and discarded vacuum cleaners that form this ghastly mountain range. So huge is the mass that bulldozers have plowed roads that run up and over and along the long summit ridge. The treads audibly crush computer monitors and lawn mowers and dollhouses as the bulldozers move along the slopes, pushing around more and more bulk while swarms of scavenging seagulls circle overhead.

And as I stand there, peering up at this rich archaeological waste heap, surrounded by bombed-out neighborhoods that stretch toward a distant island of dead skyscrapers and the scarred hulk of the Superdome, I begin to falter with exhaustion from the tens of thousands of miles I've traveled that day. I have, that same day, muddied my shoes walking the streets of Baltimore's Inner Harbor and the fabled Financial District of lower Manhattan. I've driven the endless cul-de-sacs of Houston's southern suburbs and crossed the coastal outskirts of Atlantic City and Charleston and Savannah. I've strained my neck looking up at the abandoned skyscrapers of Miami and Boston and Tampa Bay and Seattle. I've seen, that morning, every inch of America's coastline. I've seen New Orleans after Katrina, so I've seen the future everywhere.

I'm utterly spent as I finally leave the mountain range of debris along this particular coastal city, along Westend Boulevard in New Orleans, Lousiana. As a final stop, I make my way to the 17th Street Canal levee twelve blocks away. Here, on August 29, 2005, on the western edge of the Lakeview neighborhood, the levee collapsed from a terrifying wall of rising water kicked up by Katrina's winds. Black, stinking mud several feet deep now surrounds what's left of the homes that were blown apart here, as if by dynamite, when the dam burst.

I notice a man with dark features picking through the debris of a home whose edifice is half gone. The house is two hundred feet from the levee breach. The man's name is Hadi and he moved to America from Iran as a child and prospered until Katrina came, he tells me.

"My house, it had three nice bedrooms before the flood and I drove an SUV. That one down there." He points many blocks down the street to a red SUV presently wrapped around a distant tree.

Hadi is looking for heirlooms and old photos, then he'll join his wife and son in Houston, he says, having lost his job as an engineer for the city of New Orleans. "It's not safe here anymore. I'll start over in Houston."

And after that? I wonder. And again after that? Who can keep running from something as large as this? Where is the place of true safety?

As Hadi speaks, hard-hatted construction crews work noisily and nonstop just behind us. They are repairing the levee breach—now plugged with sandbags—by installing long, armoring sheets of metal. The crews use cranes and pile drivers and, I'm told, they're at it seven days a week, twelve hours a day, racing with great emergency intensity to beat the start of the next hurricane season.

7

No More Katrinas, No More Warming, No More Waiting

On the morning of September 11, 2001, just moments before hijacked planes changed the course of America, I was carefully pouring a bushel of corn into a storage bin in my living room. I had just installed a kind of pot-belly stove that thoroughly heats my home using only organically fertilized corn kernels as fuel. The stove, equipped with electronic sensors and electric fans to maximize the heat, would soon cut my winter fuel bill in half while reducing my greenhouse gas emissions from home heating by a staggering 85 percent.

So there I was on September 11, receiving an orientation from two ex-farm-boys-turned-corn-stove-dealers, when a friend called and told me to turn on the TV. Thus it is that today, like all Americans, I can recall exactly where I was and what I was doing when two fuel-filled airliners plowed into the World Trade Center and ended the lives of three thousand people. I was converting my home to corn power.

Right after the 9/11 disaster, I was amazed by the number of pundits and average Americans who, exhibiting complete sincerity, asked the question, Why do they hate us *so* much?

The answer couldn't be more obvious. Even before the second Iraq war, our U.S. military presence in the Middle East, deployed almost exclusively to protect oil interests, was costing us $50 billion a year. We had forward-based troops on Saudi Arabian land considered holy by many Muslims. We had fighter jets on massive aircraft carriers in the Persian Gulf. If Syria had forward-based troops in New Jersey and Iran had aircraft carriers in the Chesapeake Bay, we might have a beef with them, too.

September 11 happened because of oil, plain and simple. And the only long-term way to prevent future attacks is to remove from the Middle East the U.S. soldiers and armaments that are the focus of so much Muslim extremist anger. And to do this means getting off of oil.

Which is why I can proudly tell people today that my corn stove is fighting terrorism. Instead of using oil or natural gas shipped in from thousands of miles away to heat my home, I buy sustainably raised Maryland corn from a Mennonite farmer forty miles from my house. And if my farmer's feed-truck jackknifes, God forbid, on the road down to my house, what do you have? A corn spill! The birds and squirrels and ants will promptly take care of the escaped fuel.

But my original motivation for getting a corn stove was not to fight terrorism. It was to fight global warming. Eight months before 9/11, in January 2001, I was in the middle of writing my book *Bayou Farewell* about the collapsing Louisiana coastline when I read an article on the front page of the *Washington Post* that completely changed my life. There, above the fold, was news that the Intergovernmental Panel on Climate Change was now predicting a warming of between 3 and 10 degrees Fahrenheit by 2100. This was much more than the same panel had projected just five years earlier.

I knew enough about global warming even then to put down my cereal spoon and quietly panic. A rise of two or three degrees alone, I knew, would bring staggering consequences. Worse, despite the explicit new warnings from scientists, we humans were entirely capable of bringing this calamity down upon our heads in full force. I knew this with complete certainty because, at that very moment, I was deep into writing an entire book about an American coastal society that was committing suicide. I had traveled all across south Louisiana and seen firsthand the appalling subsidence and land loss. I had interviewed the shrimpers and examined the maps and read the environmental reports. Katrina was coming. She was well on her way. And yet nothing was being done to stop her. No preparations were being made. It was an insane situation; a kind of mass psychosis. But it was happening. Entire societies can, in the face of irrefutable signs of danger, commit hara-kiri. It was happening at that very moment in Louisiana and it had happened many, many times to other peoples in the past.

And now, on the front page of the *Washington Post*, was news that the entire planet was overheating and human beings were driving the disaster. And despite assurances from scientists that an urgent international push to reduce greenhouse gas emissions could avert the worst impacts, nothing substantive was being done. And so, on that January morning in 2001, staring at the *Washington Post*, I realized something with terrifying certainty: The entire world was becoming one enormous state of Louisiana. I could vividly picture the catastrophic heat waves and crop failures and sea-level rise and massive storms because I had seen, with my own eyes, the giant watery runway human beings had carefully prepared for Katrina.

And if the whole world resembled every bit the fragile coast of Louisiana, hurtling toward disaster, then I was nothing more than a happy-go-lucky Louisianian myself, doing nothing to save myself or my family. I had not lifted a finger up to that point to fight global warming in my immediate community and surrounding region. I was just hoping the planet would get lucky somehow, that the problem would simply go away, and until it did I would just pretend it wasn't happening. But the 2001 IPCC report, forecasting up to ten degrees of warming, made it really hard for me to keep pretending the Kool-Aid wasn't laced with cyanide.

Which is why, in 2001, owing to my coincidental but intimate knowledge of the shrinking marshes south of New Orleans, I became a dedicated global warming activist. The only way our species was going to avoid global suicide was if people like me—and you—ended our state of denial and passivity and raised our voices against the powerful special interests that are driving the climate chaos.

So I devoted myself, right then, to being an agent of change. I would use every means available to me to encourage people to change the way they thought about the world and the way they conducted their daily lives and the way they voted for leaders. But before I asked even one person to make even one change to fight global warming, it was critical that I make every possible change in my own life, I felt. I recalled Eleanor Roosevelt's words: "It is not fair to ask of others what you are not willing to do yourself." Which is why I eventually replaced my 1990 Geo Prism with a hybrid Toyota Prius whose EPA fuel economy rating is a lush fifty-five miles per gallon in combined city/highway driving. This huge reduction in my greenhouse gas "footprint" was matched almost pound-for-CO_2-pound when I abandoned the typical carbon-intensive American meat diet for an all-vegetarian diet featuring such delights as veggie lasagna, fish-free Japanese sushi, and the occasional veggie burger from Burger King. (For more information on diet and climate change visit www.chesapeakeclimate.org/pages/page .cfm?page_id=66.)

But it was energy use inside my home that offered the biggest, most exciting changes. I decided to be ambitious, setting as my goal in January 2001 a 50 percent reduction in carbon dioxide emissions at home. This meant cutting in half my use of natural gas and coal-fired electricity as applied to lighting, heating, cooling, cooking, recharging cell phones—*everything*.

But I quickly abandoned this lofty goal once I got down to work because, to my delight, I had set my sights way too low. In a matter of months in 2001, I was able to retrofit my ninety-year-old home with an energy system featuring corn, solar, and wind power coupled with huge efficiency gains. My carbon dioxide emissions, as a result, dropped not

50 percent but nearly 90 percent. Better yet, I was able to do all this on a tight budget, without borrowing from a rich uncle or hitting the Powerball. In fact, as you'll soon see, my switch to climate-friendly energy sources actually saves me money over time. Lots of money.

But mine is just one household. It's not the same as transforming an entire national economy, with all its bigness and complexity. Or is it? Here's what I know for sure. No one in the world can tell me American households can't make the switch to clean, efficient energy. I know they can. I know it not because someone told me it was a good idea. Not because I read about it in a book. Not because it's a nice theory.

I know it because I've already done it.

THE BUSH ADMINISTRATION, of course, views matters entirely differently. On the plus side, the president no longer dismisses global warming as a creation of "junk science." Quite the contrary. He now openly admits that climate change is a real concern and that human beings are directly implicated. At the Group of Eight economic summit in Scotland in July 2005, Bush told reporters, "I recognize that the surface of the earth is warmer and that an increase in greenhouse gases caused by humans is contributing to the problem."

In a variety of reports and other documents put out by the White House over the past six years, the administration has gone much further in officially acknowledging this same reality. To cite one example, the White House in 2004 sent to Congress a report saying the *only* likely source of the rapid warming in recent decades was the combustion of fossil fuels worldwide, and that the warming could, among other things, have very negative impacts on U.S. agriculture yields in coming years. The report was accompanied by a cover letter signed by the secretary of energy, the secretary of commerce, and the president's chief science adviser.

So there's definitely a problem. Even George Bush says there's reason for concern. Even George Bush acknowledges the *cause* of the concern. After years of precious time lost "debating" the avalanche of unimpeachable scientific evidence, we can now put all that behind us. Unfortunately, George Bush hasn't run out of doubts. He now doubts there's any real *solution* to global warming. A rapid switch to clean, efficient energy would wreck our economy, he says. Even the modest targets of the Kyoto Protocol would be economic suicide for America, he says. The protocol just isn't practical. It isn't workable. We can't do it.

Never mind that Japan, with a standard of living equal to ours, uses just over *half* the energy we use per capita. Never mind that Denmark already gets more than 20 percent of its electricity from wind power. Never mind that Brazil powers almost half its cars on fuel derived from sugarcane.

The Bush administration says *we* can't do any of this. We Americans just aren't up to the challenge. We can't hack it. This despite the fact that virtually all we've done as a nation for the past two hundred years is solve problems—in medicine, in agriculture, in electronics, in space travel. When it comes to the greatest crisis human beings have ever faced, requiring the greatest leadership and creativity and ingenuity America has to offer, our president essentially says we're so inadequate to the task that we shouldn't even try. Sure, he pays lip service to ethanol and fuel cells, but our overall national commitment continues to be the burning of lots and lots more coal and oil and natural gas from around the world. Clean energy? It's just too big a hassle.

Bush presents this view of our nation not because it's true. Not because it makes sense. Not because it's good policy. He says all this because he's an oil man. The solution to global warming means the unavoidable death of the world's oil industry. So an oil man says global warming can't be solved.

The reality is quite different, of course. The path to a stable climate is bright and hopeful and utterly life-affirming. It's also utterly doable—right now.

8

The Clean
Energy Revolution

THE BAD NEWS, AGAIN, is we don't have much time to stabilize our fragile climate. The good news is we don't *need* much time. If we commit, right away, with a wartime intensity equal to the challenge at hand, we can cut our total greenhouse gas emissions 50 percent or more in just fifteen years or less while simultaneously improving our economy and way of life. If this sounds too good to be true, *please* keep reading.

There are specific steps we'll need to take along the way, of course. But first let's examine what we *don't* need to do. We don't need to wait around for the "promise" of hydrogen fuel cells to arrive thirty or forty years from now to solve all our problems. Fuel-cell technology will almost certainly play a huge role in our clean-energy future, and it deserves a serious national commitment to research and development right now. But the technology is decades from mass application when we need big, big emissions reductions right now. Also forget about the newly resurrected hype about nuclear power as our "clean energy" savior. Nuclear plants are dangerous and wildly expensive and they'll take too long to build. We don't have the money or the time.

Think, instead, of simpler, safer, faster, cheaper solutions. Imagine every household in America, just three years from now, illuminated each night using 66 percent less electricity. Think of every car, bus, and truck cruising down our highways fifteen years from now using 85 percent less gasoline or no gasoline at all. Think of every kitchen refrigerator using the energy equivalent of a fifty-watt incandescent lightbulb in time for the 2012 presidential election.

Yes, we're going to need lots of wind farms and solar roofs before we're done fixing global warming. But our nation's most valuable fuel on the path to a safe climate—the fuel easiest to exploit in the shortest amount of time—is our "efficiency fuel." In 2001, when Vice President Dick Cheney famously described energy conservation as "a sign of personal virtue" but not something upon which to base a national energy policy, he might as well have been discussing evil weapons of mass destruction in Iraq. The accuracy level was just as high.

The truth is this: Doing radically more work using radically less energy has to be the heart and soul of our climate rescue plan. Without sacrificing one ounce of comfort, America can dramatically cut its consumption of oil, coal, and natural gas not in a matter of decades but in a matter of years or even months. Along the way we'll save lots and lots of dollars.

Here's a real-life, gimmick-free example of just how far efficiency can carry our nation once we make up our minds to be carried. In early 2001, the world's seventh-largest economy—California—was mired in an enormous energy crisis. Energy traders at Enron and other companies had brazenly orchestrated an electricity "shortage" that was forcing rolling blackouts, driving prices sky high, and threatening to bankrupt the state's utilities. With few available options, California's politicians began pleading with citizens to use less electricity, the hope being that a drop of even a few percentage points in demand would take some of the edge off the crisis. Instead, the conservation efforts that followed were so extraordinary they practically solved the crisis on their own.

In a matter of weeks, between January and May 2001, aggregate electricity use in California dropped an amazing 11 *percent*. This saved the state at least a billion dollars while the economy continued to plug along, producing a near-record volume of goods and services on much less energy. Businesses and homeowners simply turned off unused lights and appliances. They turned down overly cold refrigerators and activated computer screen savers. They took steps that were obvious and painless and yet hugely effective simply because they were asked to. More formal sticks and carrots would be imposed by the state on energy use by early summer. But in the winter and early spring of 2001, Californians weren't acting on the promise of government rebates or the fear of legal reprisals for failing to act. They changed their behavior because it was simply the right thing to do. And it was also easy. So a staggering 11 percent fall in electricity use happened more or less on the fly, practically as an afterthought, equal to 253 million megawatt hours of power in a state representing 12 *percent* of the entire U.S. population.

What in the world would happen, I wonder, if we got *serious?* What if, as a nation, we used our airwaves and billboards to encourage similar behavior change all the time, from coast to coast, while simultaneously adopting improved efficiency standards for new appliances and cars and offering aggressive tax incentives for people and businesses who simply used less?

What would happen? Here's another small taste. In 2001, the National Academy of Sciences declared that American passenger vehicles were so ripe for improvement that Detroit could nearly double the average fuel economy of the U.S. fleet to roughly 40 miles per gallon while using existing technology. This improvement would not compromise safety or increase costs to consumers, according to the NAS study. It would also improve national security and fight global warming.

And that's just the tip of the iceberg. If, over the next ten years, we replaced every American vehicle now on the road with gas-electric hybrids similar to the Toyota Prius I drive today, our national gas consumption would be cut *in half*. This alone would reduce total U.S. carbon dioxide emissions 25 percent and virtually guarantee no more September 11 attacks or the need for future U.S. wars in the Middle East. Again, no magic bullet is needed to make this happen. The technology is sitting in my driveway right now.

Let's return to electricity for a moment. Few Americans realize that over 50 percent of our electric power comes from burning coal. Another 19 percent comes from burning natural gas and oil. As a result, nearly 40 percent of America's *total* CO_2 emissions come from power plants. Yet these plants are wildly inefficient. On average, only 34 percent of the fuel they burn is actually converted into electricity. The rest, 66 percent, is simply lost as waste heat. Indeed, according to noted energy expert Amory Lovins, America throws away enough valuable heat at power stations each year equal to the total energy use of Japan, the world's second-largest economy.

Denmark, meanwhile, generates about two-fifths of its electricity from "cogeneration" plants that actually capture the waste heat for additional production, thus converting 61 percent of its power plant fuel into electricity (versus, again, 34 percent in the United States). And one American company, Trigen, now manufactures a turbine system so efficient it converts 90 to 91 percent of a fuel's energy content into electricity. "Fully adopting just this one [power plant efficiency] innovation wherever feasible would reduce America's *total* CO_2 emissions by about 23 percent," says Lovins.

So what's the holdup? If efficiency makes such good sense for our economy and the environment, if the technology is available and consumers save money, why are we still a nation of profligate energy hogs? Much of the answer is relatively straightforward. When a carmaker or refrigerator manufacturer sells you its product, the transaction pretty much ends right there. The company won't be around to help pay the gasoline or electric bill over time. The goal of the manufacturer is to outsell the competition by bringing the cheapest, most durable product to the market. After the sale, the energy needed to make the product run is your problem.

Nor should we expect individual companies to take it upon themselves, one by one, to solve our nation's energy problems. That's the job of good public policy. And good policy is exactly what's missing in America in terms of energy waste and global warming.

Indeed, our current system often *encourages* businesses to waste energy. Electrical wiring in our homes and offices is a classic example. As Paul Roberts writes in his book *The End of Oil*, "Contractors know that thicker-gauge copper wire conducts electricity more efficiently, with less energy lost through wasted heat, than does thin wire. The difference is big enough that using a thicker wire, though more expensive to install than a thinner wire, will pay for itself through lower energy bills in less than five months. Nevertheless, contractors rarely use thicker wires because electrical work is usually done by the low bidder, whose goal, not surprisingly, is to minimize up-front materials costs, and who doesn't care about the 'life cycle,' or the operating costs of the building. The thinnest wire allowed by law is invariably installed, and the house or business essentially throws away much of the electricity before it reaches a lightbulb or appliance."

Lovins calls this business practice the CATNAP approach: Cheapest Available Technology, Narrowly Avoiding Prosecution. Businesses do this, quite frankly, because there's nothing stopping them. Absent rational standards that protect both individual consumers and society as a whole, the system gravitates toward waste.

And the waste is everywhere. Our lack of a coherent national energy policy means that barely a quarter of the energy from our cooking stoves actually reaches our food. It means only 15 percent of the energy in a gallon of gas ever reaches the tires of a U.S. car. It means that even when the electricity finally reaches the typical household lightbulb, the bulb itself creates twice as much waste heat as light. It means that Americans send to the landfill enough aluminum to replace our entire commercial aircraft fleet every three months. (This, by the way, forces us to create 60 percent of our new aluminum from virgin ore, a process that uses twenty times more energy than aluminum from recycled stock, according to Lovins.)

The tragedy is that all of this could be avoided. Setting higher efficiency standards requires of consumers only that they pay slightly more up front in exchange for lots more money saved over a product's life cycle. Switching entirely to EPA-rated "Energy Star" appliances (10 to 50 percent more efficient on average) over the next fifteen years would alone save the nation $100 billion. As Lovins likes to say, energy efficiency is not a free lunch. It's a lunch we're *paid* to eat.

Still not convinced? Still fighting the suspicion that this is all too good to be true? If so, there's not much else I can do except ask you to forget all the think-tank statistics for a moment and just let me tell you my own efficiency story. Let me tell you about my house.

IN 2001, WHEN THE SHOCKING NEW PROJECTIONS about global warming spurred me to change my life completely, I purchased a book called *Homemade Money*. Published by the Rocky Mountain Institute for people wanting to save money through improved energy use, this book was like putting on a pair of night-vision goggles. Finally I saw what had previously been invisible to me: the nonstop energy loss in every corner of my house.

As a first step, I rushed out and bought twenty compact fluorescent lightbulbs. Compared to my wasteful incandescent bulbs, whose super-hot glow was as good at burning the hands of young children as it was at lighting a room, the fluorescent lights were small miracles. Each new bulb used 66 percent less electricity, lasted ten times longer, put out an agreeable warm light, and, over its life span, would keep 260 pounds of carbon dioxide out of the atmosphere while saving me sixty dollars—*per bulb!*

Compact fluorescents are routinely on sale these days for as low as $3 per bulb, and switching out every bulb in my house took about twenty minutes.

Next, I focused my attention on my refrigerator, typically a home's biggest energy user. Employing a handy little device called a "watt meter," I quickly determined that my ten-year-old, 16-cubic-foot Sears Kenmore was using a hefty 1420 kilowatt hours per year. So I went on line and quickly found a model so energy efficient it was .5 cubic feet *bigger* than my old fridge but used only 415 kilowatt hours per year, a whopping 1000 kilowatt-hours drop. This baby sipped electricity, providing all the freezer services and refrigerator space of my old model while using the electricity equivalent of a fifty-watt incandescent lightbulb.

With all this it must have been designed for the Space Shuttle and cost me $10,000, right? Wrong. This fridge, which I immediately purchased, was itself a Sears Kenmore. But it was an "Energy Star" version, year 2001. It cost $750, just $150 more than the standard model of its size. Yet the moment I plugged it in it began saving me over $100 per year compared to my old model. So I would recoup my "premium" in eighteen months and earn back the entire cost of the fridge in seven and a half years. After that, it would be pure savings while I do my bit for a stable climate.

Nationwide, refrigerators account for a full one-sixth of household electricity use. Imagine if everyone in America took this one step, perhaps with the help of special "efficiency" loans paid back at $100 per year, the exact rate of the energy savings.

The next battle in my home energy revolution was addressing the needless waste of electricity through so-called vampire losses. These are the dozens of different appliances and gadgets in a home that suck energy dollars out of your wallet even when they're not being used. These include the cable TV box that's warm to the touch even when the TV's off. It's the red, glowing sensor light on your stereo waiting for the remote control you never use. It's the fax machine that's on twenty-four hours a day even though you only get two faxes a month, and the computer printer that you never, ever turn off.

My house had all of the above and more. Collectively, vampire losses like these account for an astounding 4 percent of all electricity consumed in America. That's 121 million megawatt hours, enough to power the entire states of Massachusetts, New Jersey, and Delaware at a cost of $1 billion.

So what to do? In my case, I simply marched out and for a grand total of $36, bought six "power strips" to accommodate all of my offending machines. Each strip features multiple electricity sockets grouped together on a panel that can itself be plugged into the wall using just one cord. I plug in a strip and then plug the cords of up to six appliances into the strip, which has an on/off switch. In my living room, for example, I plug my TV, VCR, cable box, and stereo into one power strip. When I'm using any of the above, I turn the strip on. When I've finished using the appliance, I turn the strip off and suddenly in my living room there's no flashing VCR clock, no warm cable box, no red sensor lights anywhere. The round-the-clock electricity "leakage" has stopped. It's the same throughout my house. Fax machine, printer, computer— they're all plugged into my $36 network of nifty power strips, shrinking my electricity bill still more.

The last major step in this home energy transformation involved simple behavior change. The typical American household shines like a giant torch each night, lit up from the basement rec room to the top-floor bedrooms, even though the whole family might be in a single room watching TV or in the dining room having supper. This sort of thing no longer happens at my house. My son and I and all our visiting friends turn off the lights whenever we're the last ones out of a room. Period.

The final energy change I made was probably the easiest. On nice days I simply dry my laundry outdoors, in the backyard. By avoiding the heat and friction of the basement dryer, our clothes actually last longer to boot.

And that's it. These four changes—new lightbulbs, new fridge, power strips, and modified behavior—constitute the full package of our energy reduction plan. The changes were all implemented rapidly, in the spring and summer of 2001. And the result? My household electricity bill plummeted 52 percent virtually overnight. It went from a total of 3760 kilowatt hours used in the year 2000 to an annual average of about 1800 kilowatt hours every year since.

This figure, 1800 kilowatt hours, is less than one-fifth the amount used in a typical American household each year, a reduction achieved without in any way sacrificing modern comfort. True, my ninety-year-old house has never had the power burden of central air-conditioning. But we don't need it. In the summer we get by fine with ceiling fans, a sleeping porch, and big shade trees that cool the house.

Other than this, though, my house is pretty typical. We use electricity every day, in lots of different ways—fridge, lamps, microwave, stereos, recharging cell phones, iPod. But unlike most Americans, we just don't waste it. We don't pour energy dollars down the toilet. Not that anyone would notice, mind you. Not a single visitor to my home in five years—not one—has ever recognized anything different about my Sears Kenmore or my lights or the temperature of my cable box when its idle.

So of course I understand how California shed 11 percent of its electricity load in just a few weeks without really trying. Of course I understand how western Europeans use half the energy per capita that we do while enjoying the same level of wealth. Of course I understand the studies that say our cars could easily go twice as far on a gallon of gas.

What I don't understand is what we're waiting for.

FOR ANYONE who fully accepts the reality of global warming and drops all denial about what's causing it, daily life in America can be a difficult thing to take in indeed. Seeing the mammoth, inefficient SUVs and the hand-scorching incandescent lightbulbs at every turn, and knowing the moral and practical implications of both, is like having to watch all the world's bread swept into a giant pile each day and set on fire, or watching all the world's milk trucks dump their cargo onto the street.

Deepening this pain is the waste of human lives in tandem with the wasted energy, and indeed because of it. It's the twenty-year-old American boys blown up in Iraq every day along with scores of noncombatant men, women, and children. It's the three thousand American lives *wasted* on 9/11.

But there's still more to bear. There's the daily waste of clean, renewable energy all around us: the unused power of sunlight falling on our roofs and wind power blowing along our mountains and biofuels waiting for harvest across our farms. That we have barely even begun to exploit these resources represents a huge waste of precious time and bounty.

But there are signs things are changing. In order to rapidly cut our greenhouse gas emissions by the 70 to 80 percent figure scientists say is needed to avoid the worst impacts of global warming, two things must happen simultaneously. One, as we've just seen, efficiency must undergo radical improvement—in transportation, in our homes, in industry. And two, as our energy load shrinks to a fraction of its former level, we must meet as much of that reduced load as possible with clean, renewable energy.

And here the news is again bright. At every stage of our nation's history, we have been embarrassingly blessed with great natural resources to serve our development needs. Rivers for transportation, forests for timber, minerals for industry. And now, as we face the new challenge and opportunities of a "carbon constrained" century, as clean energy soon becomes the one raw material we cannot live without, we find ourselves once again endowed with enormous resources ready for harvest.

Take wind power. The Great Plains region of the United States is so naturally gusty it's routinely referred to worldwide as the Persian Gulf of wind power. Just three states—North Dakota, Kansas, and Texas—have enough harnessable wind power to meet all of America's electricity needs.

Creating the technology to capture that power at affordable prices has been the challenge for years. But thanks to policy incentives first launched in California in the 1970s and maximized in western Europe in the 1990s, the global wind industry has achieved nothing short of a miracle in the past twenty-five years. Today, owing to improved control systems and larger turbines, a single two-megawatt windmill can by itself generate enough power for up to a thousand homes at a cost as low as four cents per kilowatt hour, competitive with natural gas-fired electricity. Back in the 1980s, wind power cost ten *times* this much.

As a result, wind is the fastest-growing energy resource in the world today, expanding tenfold to 47,000 megawatts in the last decade alone. In one year—2003—power from new wind farms grew 25 percent in the United States. In addition to cleaning our air, wind farms are an economic boon to rural farmers and other landowners who receive between $2,000 and $5,000 per year for each half-acre plot occupied by a modern windmill. (And contrary to the claims of some critics, properly sited windmills pose *zero* harm to bird populations.) There's even discussion of placing wind turbines atop abandoned oil platforms in the Gulf of Mexico, a powerful symbol if ever there was one.

The only problem is we're starting at a baseline so low it's practically zero. Despite phenomenal growth in recent years, wind power still accounts for only .1 percent of total U.S. electricity and 1 percent of global production. What's needed are government policies to move the market along even faster. In America, a good start would be ending the $25 to 30 billion in annual government subsidies to the fossil fuel industry and transferring it to renewable energy.

Solar power also has enormous potential. Every minute of every day enough energy from the sun strikes the earth equal to all the energy used by humans in a year. And every year we get a lot better at harvesting that power. The technology has advanced so rapidly that since 1970 the cost of a kilowatt hour of solar power has dropped a full 95 percent and it will likely drop another 75 percent in the next decade. The bad news is that at roughly twenty-five cents per kilowatt hour, solar power is several times more expensive than conventional electricity and therefore not yet ready for commercial-scale production the way wind is. But every time demand for solar panels doubles, the price drops another 20 percent. And thankfully demand continues to grow. Today, over 100,000 American homeowners—and I'm one of them—have installed rooftop photovoltaic (PV) panels or, even more cost effective, solar hot water systems. Even in the developing world, including in some of the poorest nations on earth, a million village homes now enjoy at least modest electrical power from these blue and glassy-sleek rectangles.

But could wind and solar power one day provide all of our electricity? Probably not. Or at least not any time soon. That's because these are so-called intermittent sources—the wind isn't always blowing and the sun isn't always shining. This causes interruptions in supply. A stable electrical grid can accommodate only so much energy of this sort, probably around 25 percent of the total load. Until an effective means is developed to store utility-scale solar and, especially, wind power (perhaps in the form of hydrogen gas via a process called electrolysis), then the rest of the electrical load will almost certainly have to come from "base load" plants powered by nukes or fossil fuels.

In terms of greenhouse gas reductions, however, this is not a deal-buster. Despite the many negatives of nuclear energy, one positive is that it generates almost no carbon dioxide. I don't advocate building a single new nuclear power plant, but neither do I advocate shutting down existing ones in the face of rapid global warming.

To repeat the discouraging state of present affairs: We get barely .1 percent of our electricity from wind and solar. But we get a full 7 percent from hydropower, landfill gas, and other climate-friendly sources. We get an additional 20 percent from nuclear power. Now, for a moment, follow the numbers. Assume we take the imminently doable step of cutting our total electricity use in half through efficiency gains while adopting national policies that overwhelmingly discourage the use of carbon-based electricity. Assuming we don't close any nuclear power plants in the process, then the nuclear portion of our total energy jumps to 40 percent and the hydro–landfill gas wedge jumps to 14 percent. If we then bump up solar and wind to the maximum 25 percent of this new energy pie, then suddenly 79 percent of our electricity is coming from noncarbon fuels. We've effectively squeezed three billion megawatt hours of fossil fuel–generated electricity out of the grid. That achieves a carbon dioxide reduction from the electricity sector of roughly 70 percent. That's an outstanding number on our way to a stable climate. And as our economy continues to grow, we can offset rising annual energy demand with ongoing new gains in efficiency combined with such steps as "fuel-switching" from coal to less carbon-intensive natural gas whenever possible.

The final major area of renewable energy potential comes from the untapped fields of the American farm belt. The so-called biofuels of our future are less likely to provide electricity than to power our cars, vans, and trucks in a climate-safe way. Corn ethanol, the subject of much recent attention, is an alcohol-based fuel created from the distillation of corn mash. Corn ethanol has a fuel value 26 percent higher than the energy required to make it. Sugarcane ethanol, which presently supplies 40 percent of the total vehicle fuel used in Brazil, is even better. Its fuel value is a whopping eight times higher than the energy required to make it. Brazil made the massive conversion to this vehicle fuel over the last thirty years, a period when its economy simultaneously grew an average of 2.4 percent a year. Such a fuel switch, then, is clearly compatible with the rise in prosperity every country seeks.

In America, perhaps the greatest promise lies in ethanol made through "cellulosic" materials. Using special enzymes that break down the fiber of a plant's entire mass (i.e., not just the kernels as in corn ethanol), this process allows conversion to alcohol of fast-growing, high-energy crops like switch grass. The process has the potential to deliver ethanol with a fuel value that's a full five times greater than the energy needed to make it. This while using a crop, switch grass, that is native to the American Plains. In the future, we could simply grow vast fields of this perennial grass, mow it once per year, then just watch it grow back on its own. Our cars, trucks, and vans would run, more or less, on the nation's lawn clippings.

MANY AMERICANS, I realize, hear the words *promise* and *great potential* applied to novel fuels like these and respond with great skepticism. Ours is a huge and complex industrial economy requiring massive inputs of dependable, affordable energy around the clock, and you're talking about windmills and prairie grass? Are you kidding?

But like the demonstrable, real-life miracles of energy efficiency—from California's 11 percent electricity drop in a few weeks to the small wonder of my 55-mpg hybrid Prius—clean energy today is decidedly not a matter of pie in the sky. Again, like the efficiency side of the equation, I know this with great certainty in large part due to the microcosm of my own home. I've seen firsthand what renewable energy can do both on my roof and inside my rooms.

Let me set the stage by discussing a few raw numbers. At the start, in the spring of 2001, I knew that converting my home to clean, efficient energy would require that I invest a decent chunk of money up front. Being of modest income, I turned to the only source of cash I had: my house. I applied for and received a home equity loan of $7,500 and promptly went to work.

As already mentioned, my first step was to cut my electricity use in half with the help of a new fridge, twenty fluorescent lightbulbs, and six power strips. Incredibly, having won a huge part of the energy battle right out of the gate, I still had nearly 80 percent of my loan money left, allowing me to meet the remaining electricity load with as much solar and wind power as I could afford.

It was also time to address my other major fossil fuel habit: natural gas. I was burning about 1200 therms of natural gas per year to heat my home, creating 13,200 pounds of carbon dioxide. My gas furnace was relatively new and efficient and my home was well insulated. So the only path to improvement was to find a whole new heating system free of fossil fuels.

Which is why just before September 11, 2001, I spent $2,400 on a "Countryside" corn stove manufactured in Hutchinson, Minnesota. No bigger than an average TV console, the stove sits in one corner of my first-floor living room and, for the past five years, has heated nearly every inch of my sixteen-hundred-square-foot frame bungalow. The stove is painted black with gold trim, and has a glass door that allows a nice view of the dancing flames.

It's also ridiculously easy to use. The stove's side hopper holds seventy-five pounds of shelled corn, and the kernels are gradually fed into the fire pit by a rotating electric augur. Everything you would do manually with a wood-log fireplace—manually add the fuel and fan the flames and poke and stir the logs—this stove does automatically with the augur, a set of high-efficiency fans, and a motorized fuel stirrer. All I do is keep the hopper full of corn and empty the ashes every ten days or so.

I also save a pile of money. Natural gas prices, high to begin with, have increased a whopping 32 percent since September 2001 while corn prices have barely moved at all. Even paying a premium for organic fertilization, my corn costs only $130 per ton. That creates a heating bill of about $600 for the entire winter for a three-bedroom house. That's the *total* bill. Over the past five heating seasons, I've easily saved a total of $2,000. In one more year I'll recoup my full stove investment.

But as impressive as this may be, I didn't switch to corn to save money. I did it to fight global warming. And here the numbers are even more impressive. Burning corn contributes almost nothing to global warming if the corn is planted without tilling the soil—an increasingly common farm practice—and the crop is raised using organic fertilization (my farmer uses manure from his turkey farm). Like all plants, corn absorbs carbon dioxide from the air as it grows, and when I burn it later to heat my home it releases no more CO_2 than it previously soaked up on the farm. So that's a wash. Even after factoring in the fossil fuel inputs on the farm—diesel for the tractor and combine, natural gas to dry and shell the corn, diesel to transport it to my home—*the net carbon dioxide released by my stove is 85 percent less than with natural gas heating.* A senior science writer for the *Washington Post* independently confirmed this astounding reduction soon after I publicly made the claim.

Agriculture advocates, meanwhile, say American farmers can grow enough corn to meet the energy needs of millions of U.S. citizens. So it's good for farmers, good for the climate, and good for consumers like me.

Now, having cut my electricity use in half and nearly eliminated natural gas use thanks to corn, I was ready at last to tackle solar power. Here I had to be careful since solar technology can be very expensive. Searching around, I found a used solar hot-water system for a modest $1,000 (new systems start at around $5,000) and had it placed on my sunny, southeast-facing back roof. Water from the basement is pumped up onto the roof and warmed by the sun. The system performs beautifully, creating over half my hot water for the year and further trimming what little natural gas I still used.

Incredibly, I still had nearly half my home equity loan left at this point, reserved for the biggest challenge of all: incorporating solar electricity into my home revolution. Photovoltaic (PV) panels can be even pricier than hot-water systems. So I did lots of homework—online and at the library—to find every last state and federal solar grant and tax credit available to me (learn more at www.dsireusa.org). I also did a lot of grunt work, installing most of the system myself. Thanks to all of the above, I was able to afford a 1.5-kilowatt solar PV system for a net cost to me of $3,396.

Despite some afternoon shading from trees, the system works reasonably well, meeting nearly half of my already pared down power load. Visitors love to see the thirty-six sleek, blue panels spread across my back roof like so many skylights, and I love the idea of solar electricity coursing through my house.

But frankly, I adopted this feature mostly for demonstration purposes and to support this vital industry. For a premium of just $4 per month, I could have just as easily purchased all my electricity from a wind farm in West Virginia, power provided by my local utility, Pepco. In fact, even an average American household, with a bloated electricity load of 10,000 kilowatt hours per year, can get all its electricity from wind power for just twenty-one extra dollars per month on the power bill. That's without any investments in efficiency or any behavior change whatsoever. This stuff just isn't that expensive. As it is, I buy a very small amount of wind power each year to cover what little electricity I don't get from the sun and to "offset" the smidgen of natural gas I use to cook my food and heat my water on cloudy days.

At the end of this home experiment, when all was said and done, I found I had happily overshot my targets. Instead of $7,500 in expenses—my original budget—I had shelled out a tiny bit more: $7,642. And instead of cutting my carbon dioxide emissions 50 percent, I had cut them by almost 90 *percent*. The combination of efficiency, corn, and solar—all adopted in the eyeblink of six months—sent my carbon dioxide emissions plummeting from 19,488 pounds per year to just under 2,010 pounds.

The specific investments, again, were:

High-efficiency Kenmore refrigerator	$ 750
Twenty compact fluorescent lightbulbs	$ 60
Six power strips	$ 36
Corn-burning stove	$2,400
Solar hot-water system (used)	$1,000
1.5 kilowatt photovoltaic system (partly installed by me)	$3,396
TOTAL:	$7,642

Every month, to pay for all this, I write an $87 check for the home equity loan and its modest interest rate. This amount is offset almost entirely, however, by my average monthly energy savings. Those savings work out to nearly $1,000 per year. In five more years, when I pay off the loan, that money will go directly into my pocket.

Best of all, as with the various efficiency features at my home, few people notice that all the energy in my home is coming from "biomass" or solar power or wind power. After glancing at the glassy spread of panels across my back roof and admiring the fireplace glow in the living room, everyone settles into normalcy. When you're perfectly warm in the winter and cool in the summer, when you're watching TV or listening to music with a hot cup of tea in your hand or a cold glass of lemonade, it just doesn't feel so much like the future in my house. Which is probably the coolest thing of all because it *is* the future: easy to get to, great to live in.

OF COURSE, others might insist it's just a cute little symbol, my house; a mere anecdote. But one thing I know for sure is this: Observing firsthand how clean energy works under one roof—mine—makes me fully prepared to believe the growing reports of much bigger successes out there in the larger "real" world. California cuts its electricity use 11 percent in a matter of weeks? No credibility strain for my brain. A fifth of Denmark's electricity comes from the spinning blades of windmills? Yes, of course. The Philippines gets 27 percent of its electricity from "geothermal" methods that tap underground heat? Iceland heats 93 percent of its homes the same way? No problem.

Indeed, while the United States as a nation does almost nothing to reduce its greenhouse gases, even China has made progress. An independent assessment published in the magazine *Science* in 2001 confirmed that China had cut its CO_2 emissions 7.3 percent between 1996 and 2000, even while its economy grew a whopping annual average of 8.3 percent during the same period. Steps taken included shutting down antiquated industrial plants, switching fuel use from coal to natural gas wherever feasible, and promoting reforestration, which absorbs carbon dioxide from the atmosphere.

Meanwhile, in my own region, the Delaware-based Dupont Company has exploded the boundaries of what's possible in industrial

America. Motivated explicitly by climate concerns, America's third-largest chemical maker has since the early 1990s cut its aggregate greenhouse gas emissions 50 percent below 1990 levels while simultaneously increasing production volume 36 percent. Improvements have included efficiency gains in all of its plants and cutting other greenhouse gases such as nitrous oxide and the harmful by-products in fluorocarbon manufacture. These production changes have simultaneously saved the company a whopping $2 billion. Dupont is not alone, either. Even oil giant British Petroleum has, since 1997, cut greenhouse gas emissions from company operations 10 percent below 1990 levels, again while saving millions of dollars.

But in my mind the most impressive success story comes from Portland, Oregon, a city that in 1993 became the first local government in America to adopt a plan to fight global warming. On a per capita basis, greenhouse gas emissions in Portland have fallen an impressive 12.5 percent since 1993 despite rapid population growth and a robust economic expansion that's created 200,000 new jobs. Measured in terms of aggregate emissions, Portland has cut its carbon dioxide budget to a level just slightly higher than what it was in 1990, while total U.S. emissions have grown a staggering 15.8 percent since then.

Often called "America's environmental laboratory," Portland took a variety of practical steps to achieve these reductions, including boosting public transit use 75 percent, powering 10 percent of city-owned buildings with "green" electricity, achieving a spectacular recycling rate of 54 percent, and planting nearly a million trees and shrubs just since 1996. Simply converting its traffic lights to highly efficient light-emitting diodes (LED lights) saves the city nearly five million kilowatt hours in electricity each year and more than half a million dollars in energy and upkeep costs.

Frankly, Portland could dramatically deepen these cuts if it weren't so hamstrung by the federal government's intransigence. A full 40 percent of Portland's CO_2 emissions come from the cars, vans, and trucks moving about the city. But only the U.S. Congress has the power to set fuel economy standards for American cars, and it steadfastly refuses to increase efficiency even a few percentage points, much less double the miles-per-gallon standard in a way that virtually all observers, including ExxonMobil, agree is feasible.

Again, if every U.S. vehicle had the same gas-electric hybrid technology as my Toyota Prius, gas consumption would be cut in half. A few additional improvements could actually eliminate gas use entirely in America. That's right, *entirely*. Simply adding a supplemental car battery, rechargeable at home with a convenient plug, and building vehicle bodies from a lightweight plastic polymer or carbon-fiber composite just as strong and safe as metal, would reduce gas use another 35 percent, according to the Earth Policy Institute's Lester R. Brown. The remaining gasoline required for such a car—already reduced by 85 percent—could itself be replaced with cellulosic ethanol, made from Kansas switch grass and resulting in a virtually carbon-free automobile.

Proven technology like hybrid engines and commercial wind farms, plus the looming promise of biofuels and lighter vehicle frames, are some of the building blocks that recently led the European Commission to propose a bold new energy plan for that continent. Set for adoption in late 2006, this plan calls for a realistic 20 percent cut in overall European energy use by 2020 and an increase in renewables to 12 percent of total energy use by 2010. Thanks to dedicated public policy, wind power alone in Europe is projected to provide power for 195 *million* residential customers by 2020. That's half of all European households.

Looking farther down the road, Britain, Holland, and Germany have committed their nations to overall greenhouse gas reductions of 40 to 80 percent by 2050. These reductions are in the neighborhood of what scientists say are needed to stabilize the climate. Germany, for example, plans to cut overall energy consumption 37 percent below present levels by 2050. Of the remaining 63 percent, nearly half will come from renewables with a big emphasis on solar electricity, solar hot water, and wind power.

And remember, Germany, along with most of Europe, uses half the energy per capita as the United States right now. They're *already* twice as efficient as we are, twice as clean. And yet they're pushing forward with huge additional greenhouse gas cuts while continuing to grow Europe's biggest national economy.

So when George Bush rejects Kyoto and all other clean energy commitments urged on us by the rest of the world, he's saying we in the United States can't even do as well on energy ten or twenty years from now as Europe does *today*, much less improve beyond that. We just can't do it. We can't make the changes.

I can do it at my house. Dupont and British Petroleum can do it at their facilities worldwide. Portland, Oregon, can do it. California can make big improvements. So can China. And Europe can lead the world in standard of living *and* clean energy development.

But America as a nation? We can't even start.

When, I wonder, will we finally reject these patently false and fatal excuses from our national leaders? When will we tire of this dark myth forced on us by a government for hire, a myth serving only the self-interests of a handful of mega oil and coal companies championed by George W. Bush?

Every year in America, morally and technologically, we fall farther and farther behind the rest of the world on the issue that matters most: energy. We increasingly stand out as the primitive hunter/gatherer nation we've become. With blunt tools and raw force, we roam the planet, stumbling from spot to spot, hammering the ocean floor for more oil, sucking natural gas from below more rain forests, ripping coal from the tops of our own mountains. And when we bag our prey we devour it right there on the spot, every bit of it, with no attention to the waste of our feeding frenzy, no thought of tomorrow, of planning or saving for a future even as the hunting fields grow more ravaged each year and the trek to the next kill gets harder and harder.

We do all this with a single-minded fixation even as the rest of the world turns increasingly to the very hallmark of civilization: farming. With growing finesse and intelligence, the Japanese and the Europeans and others are beginning to harvest the sun and to farm the wind and to conduct daily life with smarter, more efficient tools. And to them will soon accrue the full civilizing benefits of energy agriculture: a more stable and desirable life free of the chronic want and violence and waste and insecurity of the hunter/gatherer world, the world we Americans have made it our top priority to protect at all cost, the *American* way of life.

Let others do what they will, we announce with a great primitive roar, but no nation will tell us what to do. We'll stick to what we know. Even if it kills us.

9

Climate Cover-up
at the
White House

BUT ENOUGH with the chronic nay-saying of the U.S. fossil fuel industry and its chief mouthpiece George W. Bush. We have the capacity to transform our energy habits in a hurry. Think of it this way: Even if the very credible estimates of clean energy potential presented in this book are way off the mark—let's say by a factor of two—we're still in very good shape. If the estimates here are *twice* as optimistic as reality allows, in other words, if we can only cut gasoline use per vehicle in the next fifteen years by about 50 percent instead of nearly 100 percent and cut carbon dioxide emissions from electricity use by 35 percent instead of 70 percent, then these are still remarkable numbers that take us far down the road toward a stable climate.

We just need a national energy policy that gets us there. We need laws that set the right energy standards with the right deadlines that move creative markets in the right direction using the right technology.

But to get a national policy we need a national government that does more than simply accept that global warming is happening. It must accept the dam-burst of warnings that our changing climate poses unacceptably great danger to all of us.

And here is the crux of the problem. This president and the chief decision makers around him just don't process warnings very well. From Al-Qaeda to New Orleans to global warming, it's the same. Some weird communication breakdown seems to happen any time a messenger knocks on the president's office door and says, "Sir, there's danger coming. Clear and present danger is on the way, sir."

The pattern stretches back to the earliest days of the Bush administration. Richard A. Clarke, the veteran antiterrorism adviser who served every White House from Ronald Reagan forward, writes of the astonishing apathy toward Al-Qaeda that existed within the Bush administration prior to 9/11. In his book *Against All Enemies*, Clarke documents how his urgent and explicit requests to move terrorism up the administration's list of priorities were totally ignored despite growing evidence of a coming Al-Qaeda attack on U.S. soil.

After 9/11, of course, the administration rushed into war in Iraq despite what we now know were explicit CIA warnings that ethnic and religious conditions there could severely complicate the occupation and lead to all-out civil war.

And as far as New Orleans and its vulnerability to hurricane attack, the prestorm warnings, as we've seen, were so loud and so explicit as to seem surreal in hindsight. The full depth of this administration's inability to apprehend danger was best revealed, I think, on the day Katrina actually hit, August 29, 2005. That afternoon, a FEMA official on the scene, Marty Bahamonde, flew overhead and looked down with his own eyes, from a helicopter, at the several levee breaches ushering a torrent of water into the vulnerable city.

But Bahamonde's immediate and urgent e-mail confirming this colossal danger somehow didn't make it through Homeland Security channels and to the White House until after midnight. And then? Nothing. Deputy Homeland Security adviser Ken Rapuano later told congressional investigators that the levee breaks weren't considered confirmed because "this was just Marty's observation." Huh? So despite a detailed eyewitness report from its own emergency representative on the scene, the administration was still collecting data early Tuesday morning trying to confirm the levee breaks that much of America was by then watching on TV.

This habitual inability to accept concrete news of danger from a credible source and translate that news into early action contributed to the deaths of hundreds of New Orleanians. The fact that Bush and others were off on vacation at the peak of the crisis didn't help, of course. Bush was at his Texas ranch. Cheney was fly-fishing in Wyoming. And White House chief of staff Andrew Card and Homeland Security adviser Frances Fragos Townsend were on vacation in Maine. This despite a leaked videotape that later showed FEMA officials on August 28, the day before the storm, emphatically warning the president of impending catastrophe from Katrina.

Perhaps the Bush-Cheney response would have been better had the call gone out that Arab terrorists were systematically blowing up levees in New Orleans. Former FEMA director Michael Brown sarcastically made this observation months after the storm. And though Brown was himself fired for ineptitude after Katrina, his implicit message is well taken: The hurricane was a threat equal in scale to a major terrorist strike and thus deserved concomitant government attention. Every hour and every day that the federal response was delayed meant many more people suffered and died.

It's the same with global warming. Every month and every year that we put off the launch of a clean energy revolution means that thousands, if not millions, of people will ultimately die from future climate chaos. The U.N. already estimates that 150,000 people perish worldwide each year from the one degree of warming the world has seen so far. How many more will perish from three to ten degrees more heat?

No wonder the warnings keep getting louder almost every day in America from leaders of increasingly high stature. In January 2006, six former directors of the EPA—five of them from Republican administrations—all agreed during a gathering in Washington that global warming was a real problem and that human beings were driving it through fossil fuel combustion.

But the biggest climate warning of all since the Katrina disaster comes from faraway Greenland and its great, vanishing ice sheet. And once again, as he has many times since the great U.S. drought of 1988, NASA's James Hansen has played the role of chief messenger. In recent interviews and lectures, Hansen has emphatically stressed the fact that we may be rapidly approaching a "tipping point" toward catastrophic sea-level rise. New and extraordinary data from NASA satellites show that the Greenland ice sheet is melting at a rate of at least fifty-four cubic miles per year, twice the rate of just five years ago, Hansen says. As noted earlier, this could ultimately result in as much as *twenty-three feet* of sea-level rise. And a significant share of that rise could come as quickly as the year 2100, by some estimates.

These terrifying numbers prompted Hansen, in December 2005, to do something scientists are extremely reluctant to do outside of extraordinary circumstances. He began openly advocating for a government policy response to the data that science was bringing to the table. He declared that, by his estimation, we have just ten short years to change course as a nation and begin a serious transition to clean, efficient energy. Unless greenhouse gas emissions are stabilized by 2015 and then promptly cut, the Greenland ice sheet will likely begin a disintegration so violent and rapid that human beings will lose all ability to stop it, according to Hansen. After that, no flood gate or seawall of any kind will prevent water from completely overwhelming New York and Washington, D.C., and Tampa and Seattle.

Ten short years. Bold words from the nation's most respected climate scientist, the man who first thrust global warming into America's consciousness with his warnings a generation ago. In all the years since, Hansen's critics—including prominent "skeptics" at the Cato Institute and Competitive Enterprise Institute who count ExxonMobil among their major donors—have ridiculed him even as they themselves conveniently changed their own stories. They routinely hatched new "data" every few years to support whatever new "analysis" would support their primary political agenda: the unfettered combustion of more and more oil.

In the late 1980s, for example, the skeptics laughed and said the planet might actually be *cooling*, not warming. So burn more oil. By the mid-nineties, they said, well, it looks like the planet is warming, but humans aren't causing it and besides it's going to be good for us: longer growing seasons, more benign weather. So burn more oil. By the 2000s, in the face of new mountains of scientific evidence that even the skeptics couldn't deny, they switched yet again, adopting the current farce now mouthed by the president: The planet is warming, human beings are definitely involved, the impacts are going to hurt, but, hey, there's nothing we can do about it. The earth's climate has always been changing, and besides, clean energy just isn't something we can do. So burn more oil.

James Hansen, in stark contrast, has been utterly consistent from the 1980s forward, drawing his conclusions from carefully observed scientific data that have stood the test of time. The planet is definitely warming, he said all along, and will continue to warm, perhaps dramatically, as long as human-induced greenhouse gases continue to accumulate in the atmosphere. The past two decades show that Hansen, and thousands of other scientists within the IPCC process, have, if anything, been *under*estimating the severity of the problem and the amount of time humans have left to deal with it. Instead of the "liberal" bias the skeptics say infuses the science, Hansen and others have been very conservative in their work per the tendency of real scientists to choose cautious interpretation over even mild speculation.

All the more reason, then, to listen closely when James Hansen very publicly sets a breathtakingly short deadline: ten years. By the time my nine-year-old son graduates from high school we had better be well on our way to a wholesale energy change or *nothing* will stop Manhattan from falling twenty feet below the Atlantic Ocean.

"It would be another planet," Hansen declares.

＊   ＊   ＊

So how did the Bush administration react in December 2005 to these new satellite images of Greenland and Hansen's warning of a fast-approaching tipping point? Did it convene an emergency conference of business and government leaders to create a plan to immediately mass-produce hybrid cars and slash electricity use though feasible efficiency gains? Did it, at a minimum, rejoin the Kyoto process and work with virtually every other nation on earth to implement its modest goals?

No. The Bush administration instead asked James Hansen, its most respected climate scientist, the director of the prestigious NASA Goddard Institute for Space Studies, to please shut up. Hansen was promptly told by political appointees at NASA headquarters that, no matter what the data shows, it wasn't his job to call for urgent reductions in greenhouse gas emissions. Hansen told journalists that NASA officials ordered a review of his lectures, papers, interviews, and Internet postings and then threatened him with "dire consequences" if he continued to call for policy measures to combat climate change. Indeed, a twenty-four-year-old former Bush campaign organizer from Texas, George Deutsch, joined the NASA public affairs office and was given the explicit task of monitoring Hansen's public comments. (Deutsch later resigned in disgrace when it was discovered he had lied on his résumé.)

These shocking facts—fully substantiated by the *New York Times*, CBS's *60 Minutes*, and other media—speak for themselves. They point to the Bush administration's true "policy" on global warming. Instead of heeding the warnings and taking steps to protect you and me and our children, the Bush policy is to cover up the evidence, suppress the truth, and intimidate world-renowned scientists who attempt to serve as credible messengers to the public.

With this administration, the greatest threat of the twenty-first century long ago ceased being an honest matter of political disagreement over what to do about a certain set of immutable scientific facts. Bush, instead, has systematically attempted to *hide* the facts from the American people. And when he can't hide the facts, when he's forced to tell at least part of the truth as in his bogus 2002 "Climate Action Plan" and the 2004 letter to Congress, he then works to minimize, distort, and confuse the facts.

Soon after Bush took office, for example, American scientist Dr. Robert Watson was removed from his post as director of the prestigious Intergovernmental Panel on Climate Change. This happened because ExxonMobil, unhappy with Watson's vocal promotion of the evidence of global warming, persuaded the White House to use its diplomatic clout to oust him. In 2003, with equal audacity, White House officials removed almost all references to global warming from a landmark, six-hundred-page EPA report on the "state of the U.S. environment," leaving only a few neutral sentences on the topic (again, out of *six hundred pages*). To the same report, Bush officials *added* a prominent reference to a new study, partly financed by the American Petroleum Institute, that questioned the fast pace of recent warming.

Similarly, early in Bush's second term, former oil industry lobbyist Phil Cooney, serving as chief of staff of the White House Council on Environmental Quality, resigned after it was revealed he had inappropriately edited official reports produced by federal climate scientists. Before the global warming reports could be released, Cooney, with no scientific training whatsoever, would change the wording of key passages to create a tone that minimized the significance of the climate data and created a sense of uncertainty around various scientific conclusions where none existed.

"This is like the White House directing the secretary of labor to alter unemployment data to paint a rosy economic picture," said Jeremy Symons, a climate policy expert at the National Wildlife Federation.

When exposed, Cooney left the White House and, with apparently no concern whatsoever about appearances, went to work for ExxonMobil within days, leaving little doubt what industry was dictating federal policy on global warming.

Back at NASA, meanwhile, Hansen openly defied the government's attempt to "gag" him in 2005 and continued to speak out, reminding the White House that the very first line of NASA's official mission statement is "to understand and protect our home planet." He simultaneously referred journalists to yet another federal agency, the National Oceanic and Atmospheric Administration, where he said attempts to muzzle scientists were even more severe. "It seems more like Nazi Germany or the Soviet Union than the United States" at NOAA, Hansen said.

NOAA is widely considered the nation's leading science agency. Among its many vital offices are the National Weather Service and the National Hurricane Center, both relevant to this discussion. Yet a January 2006 *New York Times* article revealed that many climate scientists at NOAA are no longer permitted to take calls from journalists unless the interview is okayed by administration officials in Washington and is conducted with a public-affairs officer present. Top Bush appointees at both NOAA and NASA claim such media rules are simply meant to ensure that scientists, when speaking to journalists, clearly distinguish between what is scientific fact and what might be considered opinion or policy recommendations.

But the practical effect of such policies is to create an overall chill on communication between the government-funded community of climate scientists and the public. Ronald Stouffer, a climate research specialist at NOAA's Geophysical Fluid Dynamics Laboratory in Princeton, New Jersey, seemed to speak for many scientists when he estimated his requests for interviews from journalists had dropped by half because of the time it took to get "clearance" from NOAA headquarters.

But among the many government efforts—some subtle, some blatant—to cover up the climate crisis, the most conspicuous in my mind, especially in post-Katrina America, lies in the realm of hurricanes.

Despite two major studies in 2005 linking global warming to more intense hurricanes (see chapter six), NOAA's official online magazine declared in an end-of-the-year headline: "NOAA Attributes Recent

Increase in Hurricane Activity to Naturally Occurring Multi-Decadal Climate Variability." The story made no mention whatsoever of the widely publicized hurricane studies by MIT's Kerry Emanuel and others. Even after a third study in March 2006 by Georgia Tech scientists linked bigger storms to warming oceans, the NOAA Web site continued to totally ignore the tsunami of data, sticking instead to the natural cycle story as this book went to press. It's as if these landmark studies never happened.

"NOAA talks about natural cycles, but there is no evidence this is cyclic," Emanuel told a gathering of scientists at the University of Rhode Island in March 2006. He went on to pinpoint the real problem. "Scientists at NOAA have been told there is a gag order on [discussing the impact of] global warming. A U.S. government organization should not have a gag order on science. Even in Cuba, scientists can't talk about politics, but they can say anything they want about science."

WE ARE RAPIDLY EXPORTING to every coastal city in the world the basic conditions that wiped out New Orleans. This is the chief argument of this book. The two major features of that threat, the features that displaced over one million Gulf-coast residents—namely, rapid relative sea-level rise followed by a huge storm—are coming soon to a coastal city near you thanks to global warming.

Yet when Dr. James Hansen of NASA/Goddard tries to talk about sea-level rise, when he says the water could come up so fast and so furious as to imply a completely different planet, our government's *policy* is to threaten him with "dire consequences." When three major peer-reviewed studies conclusively show that Katrina-scale storms are growing more frequent worldwide, driven by the rising heat, our government's *policy* is to prevent even a single paragraph about these studies from reaching the Web site of the nation's premiere science agency.

You don't have to own a destroyed home in Lakeview or be the widow of a drowned spouse in the Lower Ninth Ward to see this policy of cover-up and obfuscation for what it is: morally sickening. Sure, political meddling and industry influence have always been a "normal" part of our government's machinery, for better or for worse.

But global warming is not a "normal" issue. It's not one more item in the tug of war between left and right with the spoils of control going to the party in power and its industry backers. The stakes are way too high this time and *none* of us can pretend otherwise any longer. No conceivable excuses can stand up to the extraordinary evidence of danger. The fossil fuel lobby's chokehold on U.S. climate policy lost all semblance of acceptability long ago. The industry's revolving-door operatives, skipping in and out of the White House to install political appointees at the highest levels of NASA and NOAA and elsewhere, these lobbyists, these men and women masquerading as public servants, are engaged in something far removed from "business as usual." As Pulitzer Prize–winning journalist Ross Gelbspan writes in his book *Boiling Point*, "This time the corruption is not leading to some unemployed workers, some defective products, or some diminished pension funds. This time it involves the future of this civilization."

The truth, then, is harsh but inescapable: All the words and deeds of the oil lobby—of the CEOs and paid "skeptics" and government apologists—have risen to the level of full-blown crimes. They are crimes not just against the American people, not just against all that we stand for as a nation. They are crimes against humanity itself, against all human beings everywhere, alive now and not yet born into that still avoidable but rapidly approaching and ultimately irreversible future world of climate chaos, violence, and death.

10

A Way Out of This Mess: Kyoto and Beyond

HERE'S WHAT GIVES ME HOPE: In the middle of my urban neighborhood, just a few miles from the U.S. Capitol, stands a tall and shiny symbol of rural America. It's a twenty-ton corn granary, full of organically fertilized, Maryland-raised corn. Every few weeks during the winter, a Mennonite farmer forty miles away dispatches a feed truck to refill this two-story-tall, cylindrical granary. Then, in the shadow of high-rise apartment buildings within earshot of D.C. subway trains, fifty families in and around my neighborhood come at their convenience to withdraw the fuel they need to heat their homes with corn-burning, climate-friendly stoves.

This first-in-the-world urban corn cooperative exists because the granary exists. And the granary exists because my city government in Takoma Park, Maryland, has a policy of fighting global warming. Indeed, the granary itself sits on city property at the public works compound. Among other things, Takoma Park's leaders recently purchased wind energy to power all government buildings. And in 2002 the city worked creatively with local citizens, farmers, and private industry to establish this unique granary system that's now being replicated in other parts of the region.

After I installed my own corn stove on September 11, 2001, I told people I was fighting terrorism. Now I tell people I'm fighting hurricanes, too. And so are many of my neighbors thanks to the outpouring of creativity and problem solving that comes when elected officials adopt innovative policies that convert common problems into public gain. The Takoma Park granary, at no cost to taxpayers whatsoever, has reduced the heating bills of lots of people. It's helped preserve at least two Maryland farms, enhanced the bottom line of a stove-manufacturing company in Minnesota, and, oh yeah, kept hundreds of tons of carbon dioxide out of the atmosphere in the process. It's also made local politicians look very, very good (more on all this in a moment).

So imagine—just *imagine*—what would happen if we did this sort of thing on a national scale. In May 1961, President John F. Kennedy committed the United States to a policy of putting a man on the moon by the end of the decade. What made that national commitment particularly audacious was that it required lots of technology that simply didn't exist in 1961. So we had a policy—go to the moon—but not the technology to get there, and still we succeeded in eight short years.

It's the opposite with global warming. We have all the technology we need to solve the problem, we just don't have a policy. We have modern wind turbines and ethanol production techniques and hybrid car technology. We just need a government that gives a damn.

Of course, the Bush administration insists it *does* have a coherent and effective plan to fight climate change. It's called volunteerism. The president, like his father and Bill Clinton before him, has called on American businesses to *voluntarily* cut greenhouse gases. But very few businesses have stepped forward for reasons we'll soon see, and carbon dioxide levels have continued to rise year after year. This approach is so absurd it's like asking American businesses to voluntarily hunt down

Osama bin Laden on behalf of the U.S. people, or voluntarily repair and maintain the New Orleans levee system. Certain threats to public safety obviously require real government commitment and leadership, *not* volunteerism.

The president also says he's addressing the climate crisis by investing billions of dollars into hydrogen fuel cell research. But this initiative is little more than a convenient escape hatch for ExxonMobil and friends, since fuel cell technology has lots and lots of problems and is at least a generation away from widespread application. But wait, there's also the billion dollars per year the White House spends on scientific research into the causes and impacts of global warming because they feel we don't know *nearly* enough to take corrective action just yet. And even when the results of all that taxpayer-funded research come back to the White House, when the NASA satellites show the Greenland ice sheet imploding, for example, we simply bury the information and tell the scientists to please be quiet. It's just one more category of waste in the great tragedy that is unchecked global warming.

But what if we had a real national plan to fight the climate crisis instead of a policy of delay and denial and cover-up? What would such a plan look like? What steps would we need to take? Are any specific plans already on the table, waiting to be put into action?

The answer to the last, thankfully, is yes. Several viable plans are ready to go. They vary from the Kyoto Protocol on the international level to several U.S.-specific plans with names ranging from the hopeful-sounding "Sky Trust" to the heroic-toned "Apollo Project for Good Jobs and Energy Independence."

But before examining some of these blueprints, it's important to note that, while Washington sleeps, a number of U.S. states have already adopted their own laws and regulations to promote clean, efficient energy. Twenty-two states, for example, including oil-rich Texas under then-governor George W. Bush, have adopted phased-in clean electricity standards that require anywhere from 3 to 25 percent of a state's electricity to come from wind, solar, biomass, or other renewable sources. Several states, including my own, have passed improved energy efficiency standards for certain appliances and tools. And soon a regional "cap and trade" system, stretching from Maine to Maryland, will ensure at least modest reductions in CO_2 emissions from power plants. Finally, several states, led by California, are now in court fighting Detroit and the Bush administration for the right to set their own standards for automobile greenhouse gas emissions.

These efforts, taken together, are not insignificant. California's actions are particularly impressive. In 2005, Republican governor Arnold Schwarzenegger committed the state to an 80 percent reduction in greenhouse gases by 2050, with aggressive cuts starting in 2010. "The debate is over," Schwarzenegger said in announcing the plan. "We know the science. We see the threat. And we know the time for action is now."

But this patchwork quilt of regions and states and even cities setting their own energy policies has created what *U.S. News and World Report* calls "an increasingly energy-schizoid land." The situation emerges unavoidably from the complete policy vacuum at the national level. That vacuum triggers laudable though uneven efforts across the land while creating unnecessary confusion and uncertainty for businesses and consumers alike.

What's needed is an end to the vacuum. What's needed is an *American* policy that enhances, unifies, and streamlines the best of these scattered internal efforts and simultaneously integrates them into a coherent global strategy to save our global climate.

THE FIRST QUESTION to ask when stepping back to view the big picture is: Who actually owns the sky? Does the government own the sky? Do corporations own the sky? Does *anyone* have legitimate claim to the atmosphere above us?

Energy visionary Peter Barnes asserts with great moral suasion that we *all* own the sky—every man, woman, and child—in equal measure. What's more, the sky is far too vital and fragile to be treated as a worthless dumping ground of harmful greenhouse gases. It simply can't hold much more of these gases while supporting life as we know it on earth. Therefore, Barnes writes in his book *Who Owns the Sky?*, fossil fuel companies must pay a kind of landfill fee—or "skyfill" if you will—that reflects the sky's scarcity of space for accepting more gases. The money would then be divided evenly among every American citizen, including children, as an annual dividend.

This so-called Sky Trust is modeled in part on the Alaska Permanent Fund, which since 1983 has distributed that state's mining and oil-drilling royalties to all Alaskans equally. In 2005, every resident got a check for $845, thus underscoring the potential popularity of a nationwide Sky Trust.

Under Barnes's plan, energy companies would be required to purchase permits applied to fuels at the moment of extraction—for example, at coal mines and oil and gas wells—and at offloading sites for foreign tankers. The cost of these permits would rise steadily over time based on government benchmarks reflecting the best scientific estimate of the climate's stabilization needs. Of course, these permits would in turn raise the price of carbon fuels for consumers, thus spurring the kind of conservation, efficiency, and the switch to renewable energy that is the end goal. When coal-fired electricity is ten cents per kilowatt hour on the wholesale market and gasoline is $5 per gallon at the pump, then money-saving wind power and low-cost ethanol fuels will pour even faster into the market.

But the beauty of the Sky Trust plan is that it ensures equity and fairness in the transition to a clean-energy economy. Again, every American—rich and poor, young and old—would receive the same annual dividend, perhaps as much as $1,000 per year at the start. But in practice, the Hummer driver would appropriately pay much more into the fund than the Prius driver through greater gas purchases. Rewards and penalties would accrue in this way throughout the economy, based on standard market principles. The poor, meanwhile, who consume less of everything, including energy, and who are most vulnerable to price increases, would be insulated from any undue burden by the annual dividend payment.

Part of the Sky Trust would also be used to fairly compensate and/or retrain coal miners and other workers in the fossil fuel business as that industry inevitably shrinks and fades away. Barnes envisions the federal government as administrator of the trust fund per the successful state-government model in Alaska.

Besides being fair, this system of tradable energy permits under an overall cap on carbon-fuel extraction—a so-called cap-and-trade system—has already been used in a different form with great success in America. George Bush senior adopted a cap-and-trade approach to fight acid rain in the early 1990s and the desired reductions in sulfur emissions from power plants have occured faster than anyone imagined possible and at a cost much lower than projected by either industry or environmentalists. When properly designed and enforced, the cap-and-trade approach works.

On balance, Sky Trust is an exceedingly elegant mechanism that has all the right ingredients for success: market efficiency, fairness, and the potential for quick results. But would Congress ever approve such a system? I think so, especially as rising seas and bigger hurricanes and other impacts of global warming intensify. One envisions a voter coalition that has as its base the lower- and middle-income voters whose dividends would far outstrip their higher energy costs. Combine this bloc with wealthier voters who understand the environmental and national security imperatives for 100-mpg SUVs, and you'd have an alliance that might break even the stranglehold of Big Oil and Peabody Coal on Washington.

Of course, if the Sky Trust proves ultimately too exotic for national acceptance, there's still the old-fashioned approach of step-by-step statutory and regulatory reform. One critical first step embraced by nearly every climate rescue plan is the immediate end to the roughly $25 to $30 billion in U.S. government subsidies given unconscionably each year to the fossil fuel industry. All of that money should be redirected, right now, to clean and efficient energy development.

One plan for energy transformation, called the Apollo Project, has received much attention in recent years. Launched by an exciting alliance of labor unions and environmental groups, the project's comprehensive ten-point plan recalls the urgency of the 1960s moon shot. Among other steps, the plan calls for the mass production of hybrid cars, the promotion of "green buildings," and ambitious incentives for solar and wind farm development. Capitalizing on multiple studies that show a clean energy economy would create many more jobs than the fossil fuel system presently does, leaders of the Apollo Project assert their plan will trigger a net gain of at least one *million* jobs in America.

A major controversy of the Apollo Project, however, is its promotion of so-called clean coal technology. The plan calls for investments in technology that would capture carbon dioxide at coal-burning power plants and pipe it to natural burial caverns deep inside mountains or inside abandoned mine shafts for permanent storage. But this process is untested and potentially highly risky as well as expensive. We're much better off, in my view, investing that money in alternatives to coal while transitioning America's fifty thousand coal miners to the new energy economy.

A central virtue of both Sky Trust and the Apollo Project is that they would create, finally, national standards under a national strategy, thus providing American businesses with the certainty and uniformity they crave and that's sorely lacking under the current state-by-state approach. Industry reflexively opposes almost all regulation, of course— Detroit opposed mandatory seat belts and air bags for years. But if the end result is a level playing field affecting all competitors equally, companies have a history of adapting in stride to appropriate regulations and continuing to prosper nicely.

But whether we adopt the Apollo approach or Sky Trust or some mix thereof, it's critical we get moving immediately. George W. Bush has delayed matters several critical years, digging our hole even deeper, and the process of debating, passing, and implementing a national climate plan, even if expedited, will flirt dangerously with James Hansen's ten-year deadline for major progress.

Thankfully, while we're doing the hard work of restructuring our economy to reduce carbon dioxide levels, there are other important greenhouse gases we can simultaneously reduce with less effort and greater speed, according to Hansen and others. Immediately placing "scrubbers" on old coal-fired power plants to reduce soot and smog-causing nitrous oxide, as my state of Maryland is doing right now, would help significantly to reduce "tropospheric," or near-surface, warming. Even more critical is the need to reduce the greenhouse gas methane, a gas far less voluminous in the atmosphere than CO_2 but twenty-one times more potent at trapping heat. Sources of methane include leaks from landfills, coal mines, and natural gas pipelines. Stopping these leaks would require comparatively modest effort and provide a great service since methane accounts for roughly 20 percent of all atmospheric warming now affecting the planet. And since methane dissipates from the atmosphere just twelve years after it's released (unlike CO_2's one-hundred-year life span), any reduction would provide the quick early results we need while we do the tougher work of ramping down CO_2.

But to really make a difference with methane, it's necessary to address the number one source of the gas worldwide: animal agriculture. Every year, livestock raised for human consumption release nearly 90 million tons of methane as part of their natural digestive processes. And unfortunately, meat consumption worldwide has risen fivefold just in the last fifty years, and continues to rise rapidly. So any nation serious about fighting global warming would patriotically encourage a reduction in animal meat consumption in favor of farm-raised fish and savory, soy-based meat substitutes. Fighting global warming, it turns out, requires more than just politicians doing the right thing in Washington. It requires all of us doing the right thing each night at dinnertime, serving food to our children that is safe in every sense of the word.

✦   ✦   ✦

BUT WHAT ABOUT THE KYOTO PROTOCOL? Where do all these proposed domestic efforts leave us in relation to that international climate initiative most Americans have heard about and which our president has so famously rejected? The reality is this: Until we create the domestic political will for our own national climate plan, there's little hope of meaningfully joining the rest of the world under the Kyoto process. The good news is that the needed political will is rapidly emerging in America, as we'll see in chapter eleven. So our reemergence on the global stage is nigh.

In the meantime, the rest of the world has moved on with Kyoto without us while anticipating our inevitable return based on two facts: (1) no country can ignore reality forever and (2) no group of countries can solve the global warming problem without the full participation of the United States.

On their own, the Kyoto nations have so far had mixed results. In its first phase, the protocol requires industrial nations to reduce their aggregate carbon dioxide emissions by 5.2 percent below 1990 levels by the year 2012. Several nations, among them Britain and Germany, have already reached or will soon reach their assigned targets, while many others, including Japan and Canada, will be hard-pressed to meet the deadline. Frankly, the total absence of the world's biggest polluter, the United States, triggers angry complaints of unfairness overseas and dampens enthusiasm for action in some nations.

But even if America joined Kyoto today we would have to really sprint to meet our previously assigned target of a 7 percent reduction in carbon dioxide emissions below our 1990 level. Indeed, with no meaningful energy reform policy in place whatsoever, our emissions have risen a depressing 19 percent since 1990, meaning we'd have to reduce our current emissions by more than 25 percent between now and 2012. This can be done (remember, California cut electricity use 11 percent in a matter of weeks in 2001), but it will take a sharp national commitment to immediate efficiency gains and behavior change while we retool Detroit and the electricity sector for the long haul.

Meanwhile, there are several flaws in the Kyoto process itself worth noting. The same cap-and-trade technique that has worked so well in reducing sulfur emissions from U.S. power plants was, at the insistence of the Clinton administration in 1997, made a cornerstone of the Kyoto mechanism for reducing CO_2. But while this technique seems to work well under the uniform conditions of a single country, it becomes extremely complicated and prone to abuse on the global scale, as Kyoto nations have discovered.

To address this shortcoming, Ross Gelbspan, a leading spokesperson for the U.S. climate movement, recommends amending Kyoto to include features of an approach called the World Energy Modernization Plan. This plan would essentially start the Kyoto clock all over again, with every country using as its baseline whatever quantity of carbon dioxide it presently emits. Then, every year after that, a nation would have to improve by 5 percent the ratio of its fossil fuel consumption compared to its total economic output (through better efficiency or a switch to renewable energy or both) until collective emissions across the planet reach the necessary 70 to 80 percent reduction. This method of uniform cuts per nation per year is much easier to verify and enforce than the complicated Kyoto cap-and-trade scheme, where, for example, Russia can take its huge store of carbon "credits," generated by the economic collapse of the old Soviet Union, and sell these "reductions" to polluting nations that don't want to make meaningful near-term pollution reductions. And the Gelbspan approach, by starting at current emissions levels, creates a more politically plausible pathway for America's reentry.

The plan also calls for the creation of a $300 billion annual fund to help bring developing nations like India and China into the process before their growing carbon emissions overtake even the industrial world. The fund, perhaps financed by a small levy on international currency transactions, would be "directed specifically to finance wind farms in India, solar assemblies in El Salvador, fuel cell factories in South Africa, and vast solar-powered hydrogen farms in the Middle East," writes Gelbspan.

This feature counters one of the Bush administration's chief criticisms—that poor nations aren't sufficiently involved in Kyoto—and uses a kind of international clean-energy Marshall Plan to help make it happen.

The ultimate truth underlying all of these details, however, is: When the United States, with all its economic and diplomatic clout as the world's only superpower, finally decides that Kyoto must succeed, then Kyoto will succeed.

There's a powerful precedent for this, after all. In 1987, against fierce objections from powerful industry constituents, the Reagan administration signed the Montreal Protocol to phase out chlorofluorocarbon emissions that were destroying the earth's stratospheric ozone layer. All the news one heard in the 1980s about global doom through ozone depletion has all but disappeared because the threat has been gradually but steadily lessened. Scientists presented hard data of impending ozone danger, a danger they said could only be addressed through global action. Politicians accepted the data and took the right steps just in time, thus avoiding a major disaster before it fully developed.

If you need a beacon of hope amid all the bleak news about global warming of late, this international triumph over ozone depletion certainly stands out. And if that's not enough, if you need more reasons to remain hopeful, there's always the admittedly obscure but decidedly inspiring story of an out-of-place corn granary, just a few miles from the White House, set up not long ago to fight the rising heat.

OF ALL MY EFFORTS, personal and professional, to protect the climate in recent years, I'm probably most proud of this: I was the first person in the entire Washington, D.C., region to use corn kernels to keep a house warm. And what began as a personal experiment in the fall of 2001, using a relatively untested technology, has now grown to a small but expanding movement with the potential to affect thousands of homes all across my region.

From the first day I turned it on, my corn stove exceeded all my expectations, conveniently heating my home and saving me money. It worked so well that a neighborhood newspaper story that first autumn inspired three other families in Takoma Park, Maryland, to purchase the devices—and a community was born.

Our immediate problem was not the technology—again, the stoves performed beautifully. Our problem was the fuel. The nearest corn farmer was a long way away, and getting our needed bushels proved harder than we first anticipated. In the end, our effort to solve this problem—to translate idealism into a workable urban infrastructure for corn fuel—triggered an amazing series of creative decisions and private-industry investments and acts of fierce political will. These actions, in my mind at least, argue powerfully for a hopeful future for our entire planet.

We four pioneering "corn families" began that first heating season driving our own cars forty miles to a farmer in Frederick County, Maryland, who was growing the sort of sustainably raised corn we sought. We would return with our front passenger seats stacked high with fifty-pound bags of bright yellow field corn. The back seats were stuffed too, of course, as well as the trunks. But fearing wear on our vehicles, we soon began borrowing sturdier pickup trucks and vans from friends, and returning with a week's worth of corn for the whole group. We even rented a U-Haul truck once and came back with several tons of the stuff.

At first this process of fetching corn was kind of exciting and romantic. We city dwellers were doing business in the bucolic countryside, getting to know a colorful farmer and his family, all because of our commitment to a better world.

But the romance wore off pretty quickly. This haphazard method of transport was time-consuming and more than a little dangerous. I recall with chills my hard-to-steer, overloaded Geo Prism rolling down narrow country roads packed to the dome light with corn. Worse, a system like this would never inspire large numbers of other urbanites to use corn for home heating, as was our ultimate vision.

"What you need is one of those," said Gary Boll, our corn farmer, one winter day as he pointed to the twenty-five-foot-tall metal granary on his land. It held feed for his turkeys and hogs. "If you could get one of those, you could store corn in bulk right down there in the city and I could send it to you twenty tons at a time and at a lower price to boot."

I was instantly sold on the idea, as were my fellow corn enthusiasts. But we had to first resolve a seemingly endless series of challenges. Foremost was this: Where in the world would we put such a granary? There was practically no open space in Takoma Park, a fully developed, century-old, inner suburb of Washington. I couldn't put it in my small backyard, not given the prospect of lots of users getting corn at all times of the day. And talk of installing it in the parking lot of the city's main grocery store—allowing people to get organic food and climate-friendly fuel in one stop!—was quickly shut down over a host of legal issues.

It became clear very quickly that the granary had to go on city property. There was no other option. And thankfully Takoma Park, a politically progressive and diverse enclave of seventeen thousand people, already had an official policy of fighting global warming. The city was a member of the Cities for Climate Protection initiative, a non-profit program aimed at helping municipalities nationwide identify, measure, and cut greenhouse gases. This meant city buildings were already highly efficient in Takoma Park and policies existed to encourage walking and biking over driving. Town leaders were also exploring the idea of buying or leasing the city's entire electrical grid—all the wires and poles—from the local utility, a move that would allow a greater use of clean electricity at a lower cost.

So when I went before the city council to explain that corn stoves for home heating could reduce carbon dioxide emissions up to 85 percent, the city leaders readily agreed to promote this method as part of their climate action plan. They voted right away to locate the corn granary at the city's public works compound on a hill behind city hall.

But the granary would cost money: $2,000 for the structure itself, another $2,000 for the concrete pad to set it on. Taxpayers could not be expected to pay this bill, city leaders said. And although this urban granary would create a badly needed new market for our farmer, he had six kids and was already struggling to hang on to his small-scale operation. He didn't have the cash to invest.

I didn't have it, either. So I sat down and composed a letter. I wrote to the president and CEO of American Energy Systems, the Hutchinson, Minnesota, company that manufactured my stove. I told Mike Haefner, a farm boy who went on to design and manufacture the first certified corn stove in America, that I loved his product. But then I explained the problem: Unless the challenge of urban storage and transport of corn was resolved, his stove market would forever be limited to the very small number of Americans living on or relatively near corn farms. The rest of America, representing at least 90 percent of all consumers, many of them willing buyers like me, would be forever out of reach.

"Imagine," I wrote, "if you were trying to sell me a really nice car at a really great price but you told me the nearest filling station was 40 miles away. That deal would fall apart. But if you told me the filling station was right in my neighborhood, right where I live, you'd sell lots of cars.

"Help us," I wrote, "set up an urban filling station for corn."

By my calculations, Mike Haefner would have to sell only four or five stoves in Takoma Park to recoup a $4,000 grant given to us. He would, of course, sell lots more stoves than that from this one granary, I told him. Besides, on the day of the ribbon cutting, he'd be on the front page of the *Washington Post*. Of that I was reasonably sure.

To his credit, Haefner immediately and enthusiastically agreed to the proposed investment. He and his nearest Maryland stove dealer would combine to cover the full cost of the granary and concrete pad.

But the deal wasn't over just yet. We had a parcel of land. We had full financing. We had a small but dedicated group of citizens ready to organize a full-blown urban corn cooperative. All we needed now, according to the city council, was insurance. We, the co-op members, had to obtain and pay for a policy that protected against damage to the granary and injury to users and any liability claims emerging thereof in this lawsuit-happy society.

"Whatever," I thought as I began what would surely be a straight-forward and relatively inexpensive endeavor. I walked into the friendly State Farm Insurance office in downtown Takoma Park thinking, well, we *are* talking about *farm* equipment.

"Run that one past me one more time," said the very cheerful underwriter after I'd laid out the situation.

"Hmmm," she said finally. "Let me get back to you on this."

She called the next day to say, sorry. A policy wasn't possible. The situation was too strange. There was no risk history upon which to base a premium. There were too many unknowns.

Undaunted, I called other insurance companies. And others. And others. And I got the same story: "What? A corn granary in the city? With big feed trucks rolling in and out and potentially hundreds of families coming and going all the time? No can do."

Even companies that specialized in insuring farms and farm equipment wouldn't touch it. The best I could do was an underwriter who said he *might* be able to get us a Lloyd's of London policy for $3,500, nearly the cost of the entire insured property itself, paid year after year after year.

This was impossible, of course, and I was starting to get a little nervous. Surely we hadn't come all this way only to have our dream snuffed out by a lack of *insurance*. I went back to the city council with a solution: What if the corn co-op accepted the stove manufacturer's money, bought the granary, had it installed on city property, then simply gave the granary to the city outright? It would become government property and could then be covered under the city's existing comprehensive property insurance policy.

"Great idea," said town mayor Kathy Porter and all the council members. "Let's do it. This should work out just fine."

Or so everyone thought until city manager Rick Finn, a longtime bureaucrat and graduate of Cyaatt College (Cover Your Ass All the Time), stepped forward to caution against the move, offering the same vague complaints of the underwriters: We've never done this before. It's too unusual. Who knows what could happen?

"Nonsense," shot back Porter, a mayor so committed to the climate fight she would soon preside over one of the only cities in America whose government buildings were powered entirely by wind energy, 100 percent. "Tell our insurer to add the granary to our policy. Tell them this is what the mayor and council want."

All seemed settled once again until a week later when the insurer, called the Local Government Insurance Trust (LGIT), shot back a letter to the mayor with objections that bordered on the outright ludicrous. The company flat-out resisted adding the granary to the city's policy. Without investigating the actual design and function of this storage device, the company said children could fall into it (no, the roof hatch would always be locked); it could spontaneously explode (no, this was for dry corn, not the methane-producing green corn, or silage, that is found in a *silo*); and, my favorite, the granary could become the target of terrorist attack! (What's next? No coverage because of a possible meteorite strike?)

The mayor's response to the company was immediate: Insure this fine granary right now, exactly as proposed, or we might consider taking our insured property worth millions and giving it to a company that truly wants our business.

In this world, when the chips are down, there is no substitute for ironclad political will. No substitute. Within a week, the insurer, LGIT, had changed its mind. It was all sweetness and light, happy to accommodate the city's proposed change. No problem whatsoever. And the added cost to the city of taking on the insurance burden of a twenty-five-foot-tall corn granary came out to, well, virtually nothing. Just $18 per year. The corn co-op immediately agreed to pay this staggering annual tab, of course.

So at last we were done. The concrete was poured, the granary was delivered, and the cooperative was officially born. We cut the ribbon on a gloriously clear and cool day in November 2002. The mayor and council were there. The U.S. congressman from our area attended. Mike Haefner came down from Minnesota wearing a big stars-and-stripes tie. And, as predicted, there was a mob of media. We were on Seattle TV and South African radio and, most satisfying, we *were* on the front page of the *Washington Post*. "A Monument to Eco-Mindedness," the headline read.

The co-op immediately tripled in size to twelve families after the unveiling of the granary. It doubled to twenty-four the next year, and is now headed for a total of more than sixty families by the end of 2006. In the meantime, we've purchased more than two hundred tons of Maryland corn at a value exceeding $25,000, all of it going to Gary Boll and his neighbor, also a farmer, whose land was brought on to help keep up with the rising demand. We've created more than $125,000 in direct sales to the corn-stove industry and generated several times more revenue for manufacturers and farmers across the country as a result of all the media attention given to corn fuel thanks to our granary and our attendant promotion efforts. Mike Haefner estimates our one group has created at least half a million dollars in business for farmers and manufacturers and their workers.

And these are gains for the *American* economy. It's amazing how patriotic clean energy is since the cheapest clean energy tends to come, by nature, from places relatively close to home. My solar electricity doesn't arrive from the Middle East. It comes from my rooftop. My wind power comes from West Virginia. And now I'm pouring Maryland corn into my Minnesota stove. It all makes me dream of the day when I can finally stop pouring Saudi oil into my Japanese Prius because Detroit has finally made a fifty-mile-per-gallon hybrid car ready to receive ethanol made from Kansas switch grass.

The Takoma Park granary is maintained by the co-op members themselves and operates entirely on an honor system. Families make a deposit of $400 to $600 to the co-op treasury at the start of each heating season. They then receive the combination to the lock for the granary trapdoor where corn withdrawals are made at members' convenience and recorded in a log book. Most members use five-gallon paint buckets with lids to store their corn, easily transporting twelve to fifteen buckets (about ten days' worth of fuel) in their cars for storage at home on porches or in backyard sheds. Few families spend more than $600 for the entire heating season.

This system has worked flawlessly for four years, and we've now collected enough membership dues to fulfill one of our core goals. We'll soon be making one or two $4,000 grants of our own to emerging cooperatives in other parts of Maryland and perhaps the District of Columbia. (These new community granaries will complement the corn supplier up in Baltimore who has already set up his own storage system for distribution to stove owners there.) The only condition for our grants is that recipient groups use sustainably raised, organically fertilized corn, and that each group make its own subsequent grant to another emerging group as membership dues allow. In this way I expect replication—slow at first—to greatly accelerate, pushed further by constantly rising natural gas prices and by intensifying fears about global warming.

As for the first granary, in Takoma Park, it's important to recognize the authentic forces that brought it into existence, that brought together people and government and creative markets. This granary was not the brainchild of some naive bureaucrat or an academic testing a pet theory. It was not funded as a "pilot project" by the U.S. Department of Agriculture or Department of Energy. The cooperative grew out of reality. It grew out of the real needs and concerns of real people intersecting with real market suppliers with real bottom lines. In blunt terms, the granary was a success because, without exception, it directly advanced the self-interest of every actor involved. Had any one of these actors—the stove manufacturer, the farmer, the city politicians, or the consumer activists—perceived that nothing meaningful was in it for them, the whole thing would have fallen apart. But one idea—corn as clean fuel—brought everyone together and helped *everyone* prosper.

Yes, the story of the Takoma Park corn granary, like the story of my clean-energy "eco-house," is a small slice of life in a bigger, more complicated world. But it's a real slice of life. It succeeded, without gimmicks, in the middle of that complicated world. And so it represents a reasonable microcosm of what we can expect all across our economy once we commit as a nation to a new energy future. Everyone will win. All self-interests will be advanced. Yes, we need the initial push from citizen activists. Yes, we need the government response in the form of effective and determined policies. But after that, all we need are businesses and average citizens who are willing to accept a more secure and prosperous life whether they've mastered the details of the big picture or not, whether they think global warming is a hoax or a nightmare in the making.

Stove manufacturer Mike Haefner is a really nice guy, after all, but he's not an environmental crusader. He drives a Cadillac SUV and his job is to make money for his company. Gary Boll is a conservative farmer who doesn't sympathize with many environmentalists. But he told me one day he sure agrees with the bumper sticker on my car that says: "No Farms, No Food." Meanwhile, I *am* an environmentalist, but one who also hates spending tons of money on pricey natural gas. And my elected city leaders want to help the environment but, like all politicians, they want to get reelected more than just about anything else. The granary made them look really good.

And so we came together, this diverse collection of actors. We stepped out of our roles—active or passive—as contributors to the problem. And we gratefully became, all of us, part of the solution.

11

The Bottom-up Solution: A Grassroots Rebellion

SEARCH CREWS continued to find dead bodies in New Orleans as this book went to press in the late spring of 2006. Two bodies were discovered April 17 buried below what used to be a house in the Lower Ninth Ward. The find brought to nearly 20 the number of badly decomposed bodies recovered in March and April alone, and brought to 1,282 the total number of deaths blamed on Katrina in Louisiana. State officials reported another 987 people still missing from the storm.

As I write this, over half the homes in New Orleans flooded during the storm are still unrepaired, their windows shattered, foundations broken, walls molding. Thousands of roofs still sit covered with the infamous blue "FEMA tarps." Millions of tons of debris choke city sidewalks and lots, undisturbed from where it all landed on August 29, 2005. No wonder two-thirds of the city's population is still in exile, unable to come home, including nearly 80 percent of the primary care physicians who once took care of the city.

The media, meanwhile, continue to focus almost exclusively on the "symptoms" and not the disease that led to the Katrina catastrophe. Yes, the Army Corps of Engineers appears to have restored the city's hurricane levees back to the Category 3 level they were supposed to have been at prior to Katrina. And, yes, the mayor has adopted an improved preparedness plan that focuses mostly on getting *everyone* out of the city when the next big storm approaches.

But even now, after all these months, virtually nothing has been done to treat the underlying cancer that is killing New Orleans. The catastrophic disappearance of coastal barrier islands and buffering marshes continues unchecked, enlarging the watery flight path that ushered Katrina into the city in the first place. Every day—day after day—the state surrenders another fifty acres of coastal land to the incoming Gulf of Mexico. And as sea-level rise from global warming accelerates, the erosion will only worsen until it simply won't matter how tall the levees are or how fast people can evacuate ahead of a storm. The city just won't be inhabitable. Period.

There's still time to save New Orleans, but not much. The Coast 2050 plan to rebuild buffering landforms between New Orleans and the Gulf is still on the table. So there's hope. Certainly other U.S. cities have recovered from near-total destruction against long odds. After the Great Chicago Fire of 1871 incinerated 18,000 buildings and left a third of the city homeless, reconstruction of the "Burnt District" started almost immediately. Within ten years, Chicago's central business district had actually doubled in size and downtown buildings sprang up taller than before. And just nine short years after the 1906 San Francisco earthquake killed 3,000 people and leveled 28,000 buildings, the city recovered entirely and played host to the 1915 World's Fair.

In New Orleans, too, many signs of recovery exist. Thousands of homes have in fact been repaired and reoccupied. Businesses and law firms that uprooted to Baton Rouge have returned. The football Saints are playing again in the repaired Superdome. And just five months after the storm, the irrepressible spirit of Mardi Gras gave rise to festivities in the city smaller than normal but just as lively with all the requisite throw beads and colorful krewes. And Jazz Fest played to tens of thousands of fans in April, converting a once-flooded horserace track into the massive party site it's always been, with zydeco favorites Terrance Simien and C. J. Chenier playing to revelers dining on crayfish étouffée and cold beer.

But these signs of renewal are anecdotal and doomed to a short life without a simultaneous plan to restore coastal wetlands *and* put a brake on the warming trend in the atmosphere.

It's also hard to imagine much permanent progress on the ground as long as FEMA is in charge of major relief efforts in New Orleans and the rest of the Gulf coast. The surge tide of incompetence that flooded the region the moment Katrina struck has never paused. Incomprehensibly, ten thousand temporary trailers still sat idle at an Arkansas airfield, awaiting shipment to displaced Gulf residents nine full months after the storm. At the same time, many evacuees were being sent by FEMA to $438-a-night hotel rooms in New York City and $375-a-night beachfront condominiums in Panama City, Florida.

Overall, at least $1 billion has been lost by FEMA through outright waste and bureaucratic screwups like these, according to the *Washington Post*. But this didn't stop the agency from trying to bill the near-bankrupt state of Louisiana for $3 billion worth of recovery "services," sending the first bill of $155.7 million right before Christmas 2005 with a thirty-day due date before interest would start accruing.

With a track record like this it's no wonder a bipartisan panel of U.S. senators in April called for the complete abolishment of FEMA. In its eight-hundred-page report entitled "Hurricane Katrina: A Nation Still Unprepared," the Senate Homeland Security and Governmental Affairs Committee said the hurricane had exposed deep flaws at every level of operation at FEMA that are "too substantial to mend." As a result, the committee recommended replacing FEMA with a National Preparedness and Response Authority, built completely from scratch from within the Homeland Security Department. "FEMA is just in shambles and beyond repair," said committee chairman Susan Collins (R-Maine).

This recommendation is an extreme but appropriate response to extreme failure. As such it serves as a model for how to handle that equally shocking but much, much larger area of national failure: energy policy. When the U.S. Department of Energy's chief function is to promote the transfer of billions of dollars of taxpayer subsidies and tax breaks to the fossil fuel industry, then, like FEMA, this office and its core policies are in "shambles and beyond repair." Our national government's entire approach to energy should be abolished and replaced with something entirely new, with the Department of Energy giving way to something like the Sky Trust Authority or the Apollo Energy Authority or simply the U.S. Department of Clean and Efficient Energy.

Again, we have all the tools we need. We have the capacity to make the switch to renewable energy in a hurry. We just need a determined national plan that addresses global warming for what it is: a war for which we have almost no time left to defend ourselves against attack.

The climate is changing so fast that in just the eight months it has taken me to write this book, I've had to repeatedly revise earlier chapters to include new studies and new disturbing data emerging about hurricane activity and the Greenland ice sheet and sea-level rise and other impacts. As this book goes to press, yet another study—this one led by a scientist at NOAA itself—has been published linking potential hurricane growth to global warming. Thomas Knutson examined sea-surface temperatures off the coast of west Africa, where many of the hurricanes that strike the United States get their start. He found that ocean temperatures have been rising there in recent decades and that this rise correlates with planetary warming. The potential result is for hurricanes to get a stronger and faster start out of the gate. NOAA's main Web site, by the way, gave this study prominent mention for exactly one day and then buried it while continuing to carry a story with the headline "NOAA Attributes Recent Increase in Hurricane Activity to Naturally Occurring Multi-Decadal Climate Variability." It's all purely normal, in other words. Still nothing to worry about.

But American insurance companies don't quite see it as "natural," and they've begun changing the way they do business as a result. Pointing to the combined effect of "climate trends" and population growth in coastal areas, several insurers are openly retreating from hurricane-prone U.S. shorelines. Indeed, the same week Knutson's NOAA study came out coupling potential hurricane growth with rising sea-surface temperatures and global warming, a risk-management company used by American insurers to estimate future hurricane damage warned of a 50 percent increase in insured hurricane losses in 2006 because of, you guessed it, "higher sea surface temperatures."

And the same front-page *Washington Post* article describing these calculations reported: "Since [Hurricane Katrina] made landfall along the Gulf Coast . . . Allstate Corp., the industry's second-largest company, has ceased writing homeowners policies in Louisiana, Florida, and coastal parts of Texas and New York state. Other firms have pulled back from the Gulf Coast to Cape Cod, notifying Florida [for example] of plans to cancel 500,000 policies."

It's hard to beat back thoughts that this could be it. This could be the beginning of that long, inevitable slide toward collapse, triggered by strange weather. It's all but impossible to do business in this modern economy without insurance. Our effort to erect a corn granary in Takoma Park, Maryland, was evidence enough of this for me. The truth is unavoidable: If we can't insure businesses and homes across long stretches of the American coastline, then most people will be forced to live and work elsewhere.

But what if elsewhere starts to look like the coastline? How do we insure inland farmers against crop losses in a world of unpredictable megadroughts and floods? What about rising tornado damage in the Midwest? Forest fires in the Rocky Mountain states? How do we engage in stable, insured business activities under such conditions?

The societal demise of Viking Greenland and Easter Island surely began with strange dislocations like these. Suddenly there were roadblocks to business as usual. A growing chaos emerged in daily life that didn't exist before but that moved steadily toward a familiar end, one repeated again and again in the course of human civilization and now chronicled by modern writers like Jared Diamond. Are we now seeing the early phase of the first-ever global "collapse"? The demise of human societies everywhere, all at the same time? The end of our ten-thousand-year experiment with civilization? As I argued at the outset of this book, we can retreat from megastorms along the Gulf coast of the United States, but there's nowhere for us to go when the whole planet gets turned upside down. There will be no mainland to receive us then; no safe shore for our retreat.

I've argued, too, throughout this book, that the story of New Orleans has many striking parallels to the story of the planet as a whole in this age of global warming, with similarities ranging from the fundamental "law of unintended consequences" to the detailed impacts of relative sea-level rise. And here's one more critical parallel: In both cases, in New Orleans and across the world, poor people get hurt the most.

We need no reminders of who endured the greatest pain in those days right after Katrina hit. With few cars and little money, the city's poor were left behind to die in grossly disproportionate numbers or to survive for days under conditions so appalling as to constitute what some observers have called "ethnic cleansing by inaction." These same victims, the most vulnerable people in the city, had no say whatsoever in the federal levee policies that triggered the catastrophic land loss along the coast. And they had no say in federal operations that resulted in the collapse of the hurricane levees that in turn destroyed their homes and livelihoods. These people did *nothing* to create the problems, in other words, yet they paid the highest price for these failures.

The same phenomenon applies to global warming. As a U.S. Peace Corps Volunteer in Africa in the mid-1980s, I lived and worked among desperately poor rural people in the Democratic Republic of the Congo. In these villages there was no discussion of clean electricity versus dirty electricity. That's because there *was* no electricity. None at all. And there were no roads or cars. No gas-fired hot-water tanks or oil-burning furnaces. And people ate minuscule amounts of meat. Indeed, beyond a few trees cut down for cooking fires, these people contributed virtually nothing to global warming. Africa as a whole, in fact, with nearly a billion people, generates only 3 percent of all the world's man-made greenhouse gases.

Yet this continent will be the first to suffer deeply as the atmosphere continues to warm. Indeed, scientists increasingly link the persistent droughts and other strange weather events already being observed across Africa to climate change. And on a continent already characterized by subsistence agriculture and rampant food insecurity, any change in precipitation patterns spells hunger for millions of people. Africa simply doesn't have the means to adapt to weather changes the way the United States and other wealthy countries might adapt in the early stages. Africa doesn't have the money or engineers to build massive irrigation systems and coastal floodgates. It can't turn up the air-conditioning when the heat waves come or dispense expensive pills when malaria spreads rapidly.

In Africa, global warming will mean lots and lots of sickness and hunger and death for people who have never stepped foot in a car. It is the world's Lower Ninth Ward, this continent. The same with Bangladesh, a country that contributes almost nothing to the problem but will lose more than 15 percent of its entire landmass to sea-level rise and a critical share of its coastal rice fields. And Bolivians who use only mules for transportation will lose their drinking water as Andean glaciers vanish. And Pacific Island nations, completely innocent, will disappear completely.

Meanwhile, the United States, with just 4.5 percent of the world's population, is directly responsible for a staggering 25 percent of all the man-made greenhouse gases now in the atmosphere. One-*quarter* of the world's total. No one disputes these numbers. And the White House of George W. Bush agrees that these very same gases are conspicuously warming the planet. And scientists tell us this warming will, in turn, disproportionately harm Bangladesh and Bolivia and all of Africa, the world's collective Ninth Ward.

So the cat couldn't be more out of the bag. We *know* what we're doing. We *know* we're harming innocent people. And yet, as a nation, we do almost nothing about it. Our FEMA-caliber energy policy, in fact, accelerates the whole process, promoting more and more warming.

In a perfect world, any nation that generated 25 percent of the globe's greenhouse gases would receive, in turn, a proportionate share of the total warming. America, in other words, would get 25 percent of all the globe's extra heat dumped right on our fifty states. If this were the case, if climate fairness prevailed, then Florida would *already* be a series of islands. Kansas would *already* be a scrub desert. Manhattan would *already* be wrecked by a series of Category 5 hurricanes.

But since we Americans can *share* this warming with the rest of the world, deflecting our fair portion away from ourselves, *sharing* it with Bangladesh and Bolivia and Africa, we show almost no concern whatsoever. We accept with barely a shrug the colossal harm we bring to blameless others.

This, I submit, is utterly unacceptable by any moral standard ever embraced by human beings anywhere. It is wrongdoing of the highest order, tantamount to exporting the horrifying refugee scenes at the Superdome and convention center to the rest of the world, creating a human rights violation without parallel. It is, in short, our greatest national shame. And it must end.

Right now.

IF THIS BOOK LEAVES but one major impression on you, I hope it is this: We have almost no time left. To avoid the worst impacts of global warming on the world's poor and on ourselves, we have just a few years to take major action, maybe only ten, as asserted by NASA's James Hansen and others. And given the great momentum built into our fossil fuel economy, with all the stubborn cultural and political attachments that connect us to this energy way of life, the task before us is very, very tall indeed. Some have likened it to taking an aircraft carrier moving at full speed and trying to turn it around on a dime. But this we must do if we are to avoid the fate of civilizations past, where societies collapsed in quick lockstep with the great ecological collapse all around them.

We can succeed at our rescue mission, but we must begin right now, and we must use all our might, and there will be no second chances. I'm reminded of a haunting quote from Dr. Martin Luther King who, nearly forty years ago, seemed to be writing with great specificity about global warming in 2006:

> We are now faced with the fact that tomorrow is today. We are confronted with the fierce urgency of now. In this unfolding conundrum of life and history there is such a thing as being too late. . . . We may cry out desperately for time to pause in her passage, but time is deaf to every plea and rushes on. Over the bleached bones and jumbled residues of numerous civilizations are written the pathetic words: "Too late."

There is but one path before us now, and there is but one person missing from that path: you. The government commitment we need to implement the technology we already have in the short time remaining will only come if we take action.

We need nothing short of a social and economic revolution in this country. The very lifeblood of our world—energy—must be altered forever. And history shows that change of this revolutionary sort almost never begins with the people at the top, with the people in political power who see the problem clearly and then act as catalysts, infecting the rest of us with the need to act. No, this is not how it happens. The average person has to demand action first. We have to raise our voices. Or, as Frederick Douglass famously said so many years ago, "Power concedes nothing without a demand. It never did and it never will."

A band of fifty common citizens in 1773 threw forty-five tons of British tea into Boston Harbor and launched what became the American Revolution. Northern ministers and newspaper editors agitated incessantly for abolition, and by 1862 the Emancipation Proclamation reordered the moral soul of this country. Housewives and seamstresses brought suffrage to American women. Black ministers and white students and janitors and lawyers and poets and workers of every stripe brought down Jim Crow and the Vietnam War in barely a decade.

And so there is but one path before us now. No one should think for a second that our government will take care of this crisis without our full, insistent participation at every step. Nor will mainstream environmental groups, headquartered inside the Beltway, lead us out of this mess. They are themselves a kind of "special interest" insufficiently broad in reach for the task at hand.

What we need is a grassroots movement as broad and wide as the nation itself, where every person is a potential recruit because climate change affects every person. It will be a movement led not by career environmentalists but by citizen activists like Ross Gelbspan (the muckraking Pulitzer winner) and Billy Parish (the Yale dropout turned climate organizer) and Sally Bingham (the San Francisco Episcopal priest who wears stoles stitched with windmills). But no matter what leaders emerge, this movement will not succeed without you.

We have one enormous advantage over societies like Easter Island and Viking Greenland. They died in isolation, utterly unaware of the similar, and therefore potentially avoidable, pitfalls that doomed civilizations before them. But we have this history in hand, and so we see the changes that we need to make right now. And from our own history of revolutionary movements, we see *how* to make those changes.

But if you're an American citizen, this much I know about you already: You're a very, very busy person. Collectively, we are surely the busiest people on earth: work, soccer, church, book club, garden club— you name it. We therefore have only so much time to give to any one good cause, and there are so many good causes out there. Likewise, we can only write so many checks.

But I implore you to recognize this truth: If you had to pick an issue—just one—among all the others out there, this is *the* one to pick.

Not only do we have a great moral responsibility to tackle global warming before it tackles us, but the solution—clean energy—has "multiplier benefits" beyond our wildest imagining. Every time you give a dollar to a group fighting global warming in your area, you're giving a dollar to fight asthma in children and mercury poisoning in pregnant women (from coal combustion). Every time you spend an hour making phone calls in support of a proposed wind farm in your region or an ethanol plant, you're spending an hour working to end mountaintop removal in Appalachia and to reduce acid rain in our rivers and to eliminate Code Red smog days in our cities and to bring all of our troops home from Iraq.

It's time to realize, in short, that nearly all of our greatest problems as a nation—in the realms of health, national security, the economy, the environment—flow directly from our national energy choices. Once we recognize this reality fully, then the revolution will begin in earnest.

But global warming trumps all the other ills by a mile. Which is why human beings all across the globe and for literally *thousands* of years to come will remember us. They will remember those of us alive right now, at the turn of the twenty-first century, because this is the moment it became irrefutably clear that the earth's temperature was spiking. They will remember us because either we did nothing to stop the rising heat, thus plunging all later generations into agricultural, ecological, and social chaos, or they'll remember us because we did the right thing, making the revolutionary switch to clean energy in time, avoiding future hunger and monster hurricanes, malaria and sea-level rise.

Throughout the hard years of the civil rights movement, Martin Luther King was repeatedly cautioned to go slow, to be sensitive of other peoples' need for gradual and more convenient change, to accept compromise and half measures with the promise of a full reward later.

But given the great moral obscenity of segregation and the great legal clarity of the U.S. Constitution, King responded with unassailable might and overpowering righteousness: Why should we wait one more day for our freedom? he asked. What man has the right to ask us to be unfree for *one more day*? We deserve our liberty now, we want it now, we demand it now!

And why, with a deteriorating sky above us and clues of past human implosions in hand, should we wait even one more day for clean energy? Why, why, why? We don't *have* a day to wait. We can't *accommodate* those who caution slowness and gradualism. The stakes are too high and the truth couldn't be clearer. We need the revolution now. We must demand it right now.

Take Action on Global Warming

To learn more about global warming and what you can do in your area to protect the climate, visit the U.S. Climate Emergency Council website at www.climateemergency.org. To learn more about hurricanes in particular and their connection to global warming, visit www.katrina nomore.org.

Bibliography

Ancelet, Barry Jean; Jay Edwards; and Glen Pitre. *Cajun Country*. Jackson: University Press of Mississippi, 1991.

Arendt, Anthony A., et al. "Rapid Wastage of Alaska Glaciers and Their Contribution to Rising Sea Level." *Science* 297, no. 5580 (July 19, 2002): 382–86.

Bloomfield, Janine. "The Potential Impacts of Global Warming on the Metropolitan East Coast: Critical Findings for the New York Metropolitan Region from the First National Assessment of the Potential Consequences of Climate Variability and Change." June 2, 2006: www.climate hotmap.org/impacts/metroeastcoast.html.

Bowman, Malcolm J., et al. "Hydrological Feasibility of Storm Surge Barriers to Protect the Metropolitan New York–New Jersey Region. New York City Department of Environmental Protection," March 2005. June 2, 2006: www.climatehotmap.org/impacts/metroeastcoast.html.

Boyle, T. C. *A Friend of the Earth*. New York: Viking, 2000.

Brown, Lester R. *Plan B 2.0: Rescuing a Planet Under Stress and a Civilization in Trouble.* New York: Norton, 2006.

Brown, Paul. *Global Warming: Can Civilization Survive?* London: Blandford, 1996.

Christianson, Gale E. *Greenhouse: The 200-Year Story of Global Warming.* New York: Penguin Books, 1999.

Claussen, Eileen, ed. *Climate Change: Science, Strategies, and Solutions.* Arlington, Va.: Pew Center on Global Climate Change, 2001.

———. "Making the Kyoto Protocol Work." *The New Democrat,* December 20, 2000. June 2, 2006: www.ppionline.org/ndol/ndol_ci.cfm?kaid=116&subid=149&contentid=2823.

Cogan, Douglas G. *Corporate Governance and Climate Change: Making the Connection.* Boston: Ceres, Inc., 2006.

Committee on the Effectiveness and Impact of Corporate Average Fuel Economy (CAFE) Standards, National Research Council. *Effectiveness and Impact of Corporate Average Fuel Economy (CAFE) Standards.* Washington, D.C.: National Academy of Sciences, 2002.

Committee on the Science of Climate Change, National Research Council. *Climate Change Science: An Analysis of Some Key Questions.* Washington, D.C.: National Academy of Sciences, 2001.

Diamond, Jared. *Collapse: How Societies Choose to Fail or Succeed.* New York: Viking Penguin, 2005.

Ehrlich, Gretel. *This Cold Heaven: Seven Seasons in Greenland.* New York: Pantheon, 2001.

Emanuel, Kerry. *Divine Wind: The History and Science of Hurricanes.* New York: Oxford University Press, 2005.

———. "Increasing Destructiveness of Tropical Cyclones over the Past 30 Years." *Nature* 436 (August 4, 2005): 686–88.

Energy Information Administration. "Emissions of Greenhouse Gases in the United States in 2004." Department of Energy 1073.2004, March 2006.

Flannery, Tim. *The Weather Makers: How Man Is Changing the Climate and What It Means for Life on Earth.* New York: Atlantic Monthly Press, 2005.

Gelbspan, Ross. *Boiling Point: How Politicians, Big Oil and Coal, Journalists, and Activists Are Fueling the Climate Crisis—and What We Can Do to Avert Disaster.* New York: Basic Books, 2004.

———. *The Heat Is On: The High Stakes Battle Over Earth's Threatened Climate.* New York: Addison-Wesley, 1997.

Gladwell, Malcolm. *The Tipping Point: How Little Things Can Make a Big Difference.* New York: Little, Brown, 2000.

Gore, Al. *Earth in the Balance: Ecology and the Human Spirit.* New York: Houghton Mifflin, 1992.

Goudie, Andrew. *The Human Impact on the Natural Environment.* Cambridge, Mass.: MIT Press, 2000.

Hallowell, Christopher. *People of the Bayou: Cajun Life in Lost America.* New York: Dutton, 1979.

Hansen, James, et al. "Global Warming in the 21st Century: An Alternative Scenario." New York: Goddard Institute for Space Studies. June 2, 2006: www.giss.nasa.gov/research/features/altscenario/.

Hawken, Paul; Amory Lovins; and L. Hunter Lovins. *Natural Capitalism: Creating the Next Industrial Revolution.* Boston: Little, Brown, 1999.

Hoffman, Peter. *Tomorrow's Energy: Hydrogen, Fuel Cells, and the Prospects for a Cleaner Planet.* Cambridge, Mass.: MIT Press, 2001.

Hunter, Robert. *Thermageddon: Countdown to 2030.* New York: Arcade Publishing, 2003.

Katz, Bruce; Matt Fellowes; and Mia Mabanta. *Katrina Index: Tracking Variables of Post-Katrina Reconstruction.* Washington, D.C.: Brookings Institution, 2005. June 2, 2006: www.brookings.edu/metro/pubs/200512_KatrinaIndex.pdf.

Knutson, Thomas R., and Robert E. Tuleya. "Impact of CO_2-Induced Warming on Simulated Hurricane Intensity and Precipitation: Sensitivity to the Choice of Climate Model and Convective Parameterization." *Journal of Climate* 17, no. 18 (September 15, 2004): 3477–95.

Kolbert, Elizabeth. *Field Notes from a Catastrophe: Man, Nature, and Climate Change.* New York: Bloomsbury USA, 2006.

Larson, Erik. *Isaac's Storm.* New York: Crown Publishers, 1999.

Linden, Eugene. *The Winds of Change: Climate, Weather, and the Destruction of Civilizations.* New York: Simon and Schuster, 2006.

Louisiana Coastal Wetlands Conservation and Restoration Task Force and the Wetlands Conservation and Restoration Authority. *Coast 2050: Toward a Sustainable Coastal Louisiana.* Baton Rouge: Louisiana Department of Natural Resources, 1998.

Lovins, Amory. "How America Can Free Itself of Oil—Profitably." *Fortune* (October 4, 2004).

———. "How Innovative Technologies, Business Strategies, and Policies Can Dramatically Enhance Energy Security and Prosperity." Invited Testimony to U.S. Senate Committee on Energy and Natural Resources Hearing on Energy Independence, SD-366, 0930–1130. March 7, 2006.

Maslin, Mark. *Global Warming: Causes, Effects, and the Future.* Stillwater, Minn.: Voyageur Press, 2002.

McDonough, William, and Michael Braungart. *Cradle to Cradle: Remaking the Way We Make Things.* New York: North Point Press, 2002.

McKibben, Bill. *The End of Nature.* New York: Random House, Inc., 1989.

McPhee, John. *The Control of Nature.* New York: Farrar Straus Giroux, 1989.

Moyer, Bill. *Doing Democracy: The MAP Model for Organizing Social Movements.* Gabriola Island, Canada: New Society Publishers, 2001.

Rabe, Barry G. *Statehouse and Greenhouse: The Emerging Politics of American Climate Change Policy.* Washington, D.C.: Brookings Institute, 2004.

Redefining Progress. *Climate Change and Extreme Weather Events: An Unequal Burden on African Americans.* Edited by Kenya Covington. Oakland, Calif.: Congressional Black Caucus Foundation, 2004.

Reiss, Bob. *The Coming Storm: Extreme Weather and Our Terrifying Future.* New York: Hyperion, 2001.

Retallack, Simon. *Climate Crisis: A Briefing for Funders.* Edited by Jon Cracknell. London: Think Publishing Limited, 2001.

Reuss, Martin. *Designing the Bayous.* Alexandria, Va.: U.S. Army Corps of Engineers, 1998.

Ridlington, Elizabeth, and Brad Heavner. *Power Plants and Global Warming:*

Impacts on Maryland and Strategies for Reducing Emissions. Baltimore: MaryPIRG Foundation, 2005.

Rignot, Eric, and Pannir Kanagaratnam. "Changes in the Velocity Structure of the Greenland Ice Sheet." *Science* 311, no. 5763 (February 17, 2006): 986–90.

Robbins, John. *The Food Revolution: How Your Diet Can Help Save Your Life and Our World.* Berkeley, Calif.: Conari Press, 2001.

Roberts, Paul. *The End of Oil: On the Edge of a Perilous New World.* Boston: Houghton Mifflin, 2004.

Rushton, William Faulkner. *The Cajuns: From Acadia to Louisiana.* New York: Farrar Straus Giroux, 1979.

Select Committee to Investigate the Preparation for and Response to Hurricane Katrina. *A Failure of Initiative: The Final Report of the Select Bipartisan Committee to Investigate the Preparation for and Response to Hurricane Katrina.* Washington, D.C.: U.S. House of Representatives, 2006.

Senate Committee on Homeland Security and Governmental Affairs. *Hurricane Katrina: A Nation Still Unprepared.* Washington, D.C.: U.S. Senate, 2006.

Tenner, Edward. *Why Things Bite Back: Technology and the Revenge of Unintended Consequences.* New York: Alfred A. Knopf, 1996.

Tidwell, Mike. *Bayou Farewell: The Rich Life and Tragic Death of Louisiana's Cajun Coast.* New York: Pantheon, 2003.

Totman, Conrad. *The Green Archipelago: Forestry in Pre-Industrial Japan.* Berkeley: University of California Press, 1989.

Velicogna, Isabella, and John Wahr. "Measurements of Time-Variable Gravity Show Mass Loss in Antarctica." *Science* 311, no. 5768 (March 2, 2006): 1754–56.

Watson, Robert T., et al. *Climate Change 2001: Third Assessment Report.* Wembley, Eng.: Intergovernmental Panel on Climate Change, September 2001.

Webster, P. J.; G. J. Holland; J. A. Curry; and H. R. Chang. "Changes in Tropical Cyclone Number, Duration, and Intensity in a Warming Environment." *Science* 309, no. 5742 (September 16, 2005): 1844–46.

Wohlforth, Charles. *The Whale and the Supercomputer: On the Northern Front of Climate Change*. New York: North Point Press, 2004.

World Conservation Union. *Mangroves Saved Lives in 2004 Tsunami Disaster*. December 19, 2005. June 2, 2006: www.iucn.org/tsunami/.

Wright, Ronald. *A Short History of Progress*. Toronto: House of Anansi Press, Inc., 2004.

Index

194   Index

About the Author

Mike Tidwell predicted in vivid detail the Katrina hurricane disaster in his 2003 book, *Bayou Farewell: The Rich Life and Tragic Death of Louisiana's Cajun Coast*. He has written five books centered on the themes of travel and nature. These include *Amazon Stranger* (detailing efforts to save the Eucadorian rain forest) and *In the Mountains of Heaven* (travels to exotic lands across the globe). Tidwell has won four Lowell Thomas awards, the highest prize in American travel journalism, and is a former grantee of the National Endowment for the Arts. His articles have appeared in many national publications. Tidwell is also founder and director of the U.S. Climate Emergency Council, based in Takoma Park, Maryland. A native of Georgia, he now lives in Maryland with his nine-year-old son, Sasha.